THE PROBLEM OF EXCITABILITY

Electrical Excitability
and Ionic Permeability
of the Nerve Membrane

THE PROBLEM OF EXCITABILITY

Electrical Excitability and Ionic Permeability of the Nerve Membrane

B. I. Khodorov

A. V. Vishnevskii Surgery Institute
Academy of Medical Sciences of the USSR
Moscow, USSR

Translated from Russian by
Basil Haigh

Translation editor
F. A. Dodge, Jr.
IBM Thomas J. Watson Research Center
Yorktown Heights, New York

PLENUM PRESS • NEW YORK AND LONDON

Library of Congress Cataloging in Publication Data

Khodorov, Boris Izrailevich.
 The problem of excitability.

 Translation of Problema vozbudimosti.
 Bibliography: p. 309-324.
 1. Excitation. (Physiology) 2. Electrophysiology. 3. Cells—Permeability. I. Title.
QP341.K4913   591.1'88   72-90339
ISBN 0-306-30593-3

The original Russian text, published by Meditsina Press in Leningrad in 1969, has been corrected by the author for the present edition. This translation is published under an agreement with Mezhdunarodnaya Kniga, the Soviet book export agency.

Проблема возбудимости
Электрическая возбудимость и ионная проницаемость клеточной мембраны
Б. И. ХОДОРОВ

PROBLEMA VOZBUDIMOSTI
Electricheskaya Vozbudimost' i Ionnaya Pronitsaemost' Kletochnoi Membrany
B. I. Khodorov

© 1974 Plenum Press, New York
A Division of Plenum Publishing Corporation
227 West 17th Street, New York, N.Y. 10011

United Kingdom edition published by Plenum Press, London
A Division of Plenum Publishing Company, Ltd.
4a Lower John Street, London W1R 3PD, England

Printed in the United States of America

In memory of my parents

S. B. Piastro and I. G. Khodorov

Foreword

The Russian edition of this book appeared in 1969 and immediately gained widespread recognition as a reference work for research workers interested in the physiology, biophysics, and pharmacology of excitable tissues. There are several reasons for the book's success.

It deals with a key problem in biology which has recently been the subject of very intensive study and it is of great interest to a wide scientific audience. Not only the fundamentals of the modern membrane theory of biopotentials, but also the vast factual material collected in the last decades by the study of the biophysical and pharmacological properties of the ionic permeability paths of the cell membrane, are described in the book in an authoritative yet readable form. Special attention is paid in the book to the systematic analysis of the consequences of the Hodgkin—Huxley mathematical theory of the nervous impulse for the problem of excitability. The relationship between the various parameters of excitability (threshold potential, threshold current, useful time), accommodation, and the action potential on the one hand, and the constants of ionic permeability of the nerve fiber membrane, on the other hand, is subjected to detailed examination in this context. To do this, the author has made extensive use not only of experimental results obtained on isolated fibers (especially single nodes of Ranvier), but also the results of his own investigations on mathematical models of excitable membranes. A chapter which deserves particular attention is that which deals with the analysis of the effect of the cable properties and geometry of excitable structures on the generation and conduction of the nervous impulse. In this section the author also makes extensive use of material obtained on mathematical models of nerve fibers. In the last chapter of the book a critical analysis is made of modern views on the molecular mechanisms of changes in ionic permeability of excitable membranes and results obtained in the study of this problem by experiments on artificial membranes are described. For this reason, although several excellent books devoted to modern ideas of mechanisms of activity of nerve fibers and cells exist, B. I. Khodorov's monograph in no way repeats them.

By comparison with the original Russian edition, the book has been substantially updated with new material obtained in the period 1969–1971. The additions to Chapter VIII, dealing with the analysis of mechanisms of rhythm transformation in geometrically and (or) functionally heterogeneous nerve conductors will be of great interest from this point of view; the reader will find information on the conduction of nervous impulses from a myelinated axon into an unmyelinated terminal and also on the mechanisms of automatic stabilization of the parameters of the membrane and spreading action potential.

Besides summarizing the results obtained by the author and his collaborators during many years of research, the book also gives an excellent survey of research on electrically excitable membranes. These features combine to make the book a valuable aid to scientists in many different specialties concerned with the problem of excitability.

I hope that the appearance of this English edition of B. I. Khodorov's book will lead to the broadening of contacts between Soviet and Western scientists conducting research in the field of the general physiology of excitable membranes.

O. B. Il'inskii

Preface

Excitability is the property of cell membranes to respond by specific changes in ionic permeability and membrane potential to the action of adequate stimuli.

By the character of their electrical response to stimulation, excitable membranes can be conventionally divided into membranes with a regenerative and membranes with a nonregenerative form of electrogenesis.

Regenerative responses are characteristic of electrically excitable membranes. In these membranes there is a mutual interaction between potential and ionic permeability: changes of potential due to an electrical (or a chemical, thermal, or mechanical) stimulus evoke changes in ionic permeability and ionic currents, and these in turn strengthen or, under certain conditions, weaken the original changes in membrane potential.

Generation of action potentials in nerve cells and nerve and muscle fibers of invertebrates and vertebrates is based on this reciprocal mechanism (Hodgkin, 1964b).

The course of electrogenesis is different in electrically inexcitable membranes, which include most postsynaptic membranes and the membranes of receptor endings and secretory cells. In these membranes, the connection between ionic permeability and potential is purely in one way: an increase in permeability induced by a stimulus (a transmitter, for example) leads to changes in membrane potential which, however, have no reciprocal effect on ionic permeability. For this reason the responses of electrically inexcitable membranes (generator potentials in receptors, excitatory and inhibitory potentials in postsynaptic membranes) cannot become regenerative in character and always exhibit a continuously graded dependence on the strength of the stimulus.

These differences in the relationships between membrane potential and ionic permeability in electrically excitable and electrically inexcitable membranes can be represented schematically as follows:

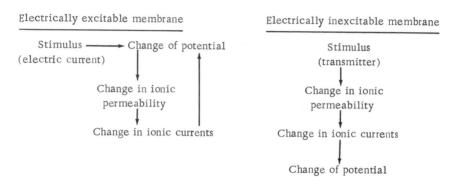

Several different types of active responses can be generated by electrically excitable membranes: local responses, graded action potentials, action potentials obeying the "all or nothing" rule, and hyperpolarization responses (see the survey by Grundfest, 1964).

A special place among these responses, because of its functional importance, is occupied by the nondecrementally propagating action potential, for it is responsible for the most rapid and perfect method of transmission of information in the multicellular organism.

Under natural conditions, the stimulus evoking generation of the action potential in a nerve or muscle fiber is the local current between the active (depolarized) and resting areas of the excitable membrane. Accordingly, the study of the principles and mechanism of action of the electric current on excitable structures has for many years been a subject for special attention on the part of investigators of many different schools and disciplines.

Particular progress has been made in the study of this problem during the last two decades as the result of advances made in cell physiology through the development and application of new techniques for use in the study of the structure (electron microscopy, x-ray structural analysis, chemical methods) and function (intracellular microelectrodes, the voltage clamp, internal perfusion of fibers, radioactive ions, and so on) of excitable membranes.

An important step in the development of the modern membrane theory of biopotentials (Hodgkin, Huxley, and Katz, 1949; Hodgkin, 1964b) was the quantitative description of the relationships between membrane potential, ionic permeability, and ionic currents in the membranes of nerve fibers. Mathematical models proposed for the membrane of the squid giant axon (Hodgkin and Huxley, 1952d) and of the single node of Ranvier (Frankenhaeuser and Huxley, 1964) not only enabled qualitative (intuitive) ideas on mechanisms of formation of the nervous impulse to be verified, but also enabled important conclusions to be drawn regarding the relationships between individual constants of ionic permeability of the membrane and parameters of excitability (threshold potential, threshold current, utilization time), of accommodation, and of the action potential.

The use of mathematical models also provided investigators with new approaches to the study of the influence of the cable properties of excitable structures on generation of the spreading action potential. Attention was particularly concentrated on the quantitative analysis of the electrical behavior of excitable tissues with complex geometry, such as the syncytia of cardiac or smooth muscle, or branching dendrites.

Detailed studies of the kinetics of ionic permeability and ionic currents during action potential generation have not yet identified the molecular mechanisms of membrane processes, and for this reason many investigations have been conducted on artificial membranes to shed light on this problem. Those receiving most attention at the present time are experiments on artificial phospholipid membranes, which under the influence of certain proteins and antibiotics acquire the properties of selective ionic permeability and electrical excitability.

An attempt is made in this monograph to summarize the main achievements of modern nerve cell physiology in the study of excitability.

The book includes the results of the author's own experimental (on single nodes of Ranvier) and theoretical (on mathematical models) investigations undertaken in conjunction with V. I. Belyaev, M. L. Bykhovskii, S. Ya. Vilenkin, E. G. Vornovitskii, R. I. Grilikhes, F. B. Gul'ko, A. D. Korotkov, E. M. Peganov, and E. N. Timin. Without their willing and productive collaboration, this book would never have been written.

Great help with the preparation of this monograph was given by O. B. Il'inskii, to whom fell the hard tasks of its reviewing and scientific editing.

Valuable critical comments have been made by M. B. Berkenblit, M. S. Markin, Yu. A. Chizmadzhev, A. M. Shkrob, and E. N. Timin.

The author is most grateful to Academician of the Academy of Medical Sciences of the USSR A. A. Vishnevskii, Director of the A. V. Vishnevskii Institute of Surgery, to his deputy, Professor A. S. Kharnas, and to Professor L. L. Shik, Head of the Department of Physiology, whose interest and support helped to transform this book from an idea to a reality.

Contents

Notation

R_{in}	Input resistance of the cell (fiber) in Ω.
C_{in}	Input capacitance in μF.
R_m	Resistance of unit area of membrane, in $\Omega \cdot cm^2$.
C_m	Capacitance of membrane per unit area, in $\mu F \cdot cm^{-2}$.
R_i	Specific resistance of axoplasm, in $\Omega \cdot cm$.
r_m	Resistance of membrane per unit length of fiber, in $\Omega \cdot cm^1$.
r_0	Resistance of external medium per unit length of fiber, in $\Omega \cdot cm^{-1}$.
r_i	Resistance of axoplasm per unit length of fiber, in $\Omega \cdot cm^{-1}$.
λ	Length constant of fiber.
E_m	Absolute potential difference between external medium and internal contents of the cell.
E	Potential on inside of membrane relative to outside potential taken as zero.
E_r	Resting value of E (resting potential).
$[\ ]_0$	External concentration of an ion.
$[\ ]_i$	Internal concentration of an ion.
V	Change in membrane potential relative to initial value E_r. A positive value of V indicates a shift of potential toward depolarization, a negative value a shift toward hyperpolarization.
$\dot{V}_{max}$	Maximal rate of rise of action potential.
E_c	Critical potential (critical level of depolarization).
V_t	Threshold potential (threshold of depolarization), i.e., the number of millivolts by which the initial value of

1

$\qquad$ E must be raised to reach the critical potential: $V_t = E_c - E_r$.

V_s $\qquad$ Amplitude of the action potential measured from the initial resting potential.

E_s $\qquad$ Value of E at the peak of the action potential.

E_K, E_{Na}, E_{Cl}, and E_l $\quad$ Equilibrium potentials for potassium, sodium, and chloride ions and of leak current respectively.

V_K, V_{Na}, V_{Cl}, and V_l $\quad$ Equilibrium potentials for the corresponding ions and leak current, measured from the resting potential E_r: $V_K = E_K - E_r$; $V_{Na} = E_{Na} - E_r$; $V_{Cl} = E_{Cl} - E_r$; $V_l = E_l - E_r$.

P_K, P_{Na}, P_{Cl} $\quad$ Coefficients of membrane permeability to potassium, sodium, and chloride ions.

P_p $\qquad$ Nonspecific ionic permeability found in medullated nerve fibers.

$\overline{P}_K$, $\overline{P}_{Na}$, $\overline{P}_p$ $\quad$ Maximum ionic permeabilities.

g_K, g_{Na}, g_l $\quad$ Conductance of membrane for potassium and sodium ions and for leak current respectively, in mhos.

$\overline{g}_K$ and $\overline{g}_{Na}$ $\quad$ Maximum values of conductances g_K and g_{Na}, in mhos.

φ_K and φ_{Na} $\quad$ Indices of reactivity of potassium and sodium activating systems to changes in membrane potential. Defined as the number of millivolts by which the membrane must be depolarized in order to increase its potassium and "peak" sodium permeabilities to half the maximum value.

$\Delta\varphi_K$ and $\Delta\varphi_{Na}$ $\quad$ Changes in φ_K and φ_{Na} relative to their initial values, in mV.

I_K, I_{Na}, I_{Cl}, I_p, I_l $\quad$ Densities of potassium, sodium, and chloride ionic currents, of nonspecific ionic current (I_p) and leak current (I_l) respectively, in $mA \cdot cm^{-2}$ or $\mu A \cdot cm^{-2}$.

I_c $\qquad$ Density of capacitance current: $I_c = C\dot{V}$.

I_i $\qquad$ Density of net ionic current, in $mA \cdot cm^{-2}$.

I_m $\qquad$ Density of total membrane current, consisting of the sum of I_c, I_i, and the externally applied current, in $mA \cdot cm^{-2}$.

I_t $\qquad$ Threshold density of stimulating current, in $mA \cdot cm^{-2}$.

I_0 $\qquad$ Rheobase density of stimulating current, in $mA \cdot cm^{-2}$.

m   Variable of activation of sodium permeability.

h   Variable of sodium inactivation.

h_∞ Stationary value of h for a given membrane potential.

$1 - h_\infty$ Index of the steady-state inactivation.

n   Variable of activation of potassium permeability.

k   Variable of potassium inactivation.

p   Variable of activation of nonspecific permeability.

α and β Rate constants of the variables m, n, h, p, and k.

$\tau_m = 1/(\alpha_m + \beta_m)$ Time constant of sodium activation.

$\tau_h = 1/(\alpha_h + \beta_h)$ Time constant of sodium inactivation.

$\tau_n = 1/(\alpha_n + \beta_n)$ Time constant of potassium activation.

$\tau = R_m \cdot C_m$ Time constant of membrane.

$\tau_s = Q/I_0$ Time constant of stimulation, where Q is the threshold quantity of electricity contained in very short pulses of stimulating current.

Chapter I

Excitability and Relationships Between the Initial and Critical Values of the Membrane Potential

The condition for generation of an action potential in a nerve fiber (or cell) is critical depolarization of the excitable membrane.

If the initial level of potential on the inside of the membrane at rest is designated E_r, and the critical value to which this potential must be raised* for a spike to develop is designated E_c, the condition for threshold stimulation can be described as follows:

$$E_r + V_t = E_c, \tag{1}$$

where V_t is the shift of membrane potential to threshold, usually known as the "threshold potential" or "threshold depolarization."

Results have been published which show that under normal conditions the value of V_t is independent both of the character of the stimulus applied and of the mode of its application to the cell.

---

*By modern convention the potential difference across the membrane is taken in the sense of internal potential minus external potential; a typical value for resting potential is −70 mV. Several traditional terms remain in common usage: d e p o l a r i z a - t i o n is a change in the membrane potential in the direction of reducing the resting potential, hence positive; h y p e r p o l a r i z a t i o n is a membrane potential more negative than the resting value; c a t h o d e and a n o d e refer to polarity of external stimulating electrodes and that fraction of the current which penetrates the fiber, acting to depolarize the membrane under the cathode and hyperpolarize it under the anode.

In motoneurons of the cat's spinal cord, for example, the action potential arises at approximately the same level of depolarization of the membrane of the initial segment both when the cell is stimulated directly and when it receives nervous impulses along orthodromic (monosynaptic, polysynaptic) pathways (Coombs, Eccles, and Fatt, 1955).

Similar relationships have been found by investigation of other nerve cells (Bennett, Crain, and Grundfest, 1959), such as the supramedullary neurons of the fish Spheroides maculatus, and also of muscle fibers (Jenerick, 1956).

Since the value of V_t is independent of the external conditions of stimulation, the threshold potential can be taken as one of a number of adequate indices of the excitability of the structure studied. Further, since

$$V_t = E_c - E_r, \tag{1a}$$

it can be concluded that the excitability of the cell is a function of the difference between the initial (E_r) and the critical (E_c) values of the membrane potential.

The closer E_r is to E_c and the lower the threshold potential, the higher the excitability of the tissue. Conversely, an increase in the difference between the levels of E_r and E_c leads to an increase in the threshold — and to a lowering of excitability.

It also follows from what has been said that the threshold potential may have the same value despite different absolute values of E_r and E_c.

Let us consider the principal types of relationships between the resting potential and critical potential observed when the nerve fiber is subjected to various influences.

Changes in V_t Associated with Primary Shifts of E_r

"Physiological Electrotonus"

Over a century ago, Pflüger (1859) showed that if a steady current passes through a nerve the thresholds of stimulation are

lowered in the region of the cathode and raised in the region of the
anode. The work of Werigo (1883, 1888, 1901a,b) added important
new facts to Pflüger's discovery. He found that the initial catelectrotonic increase in excitability is gradually replaced, if the
current continues to act, by cathodic depression, and the decrease
in excitability in the region of anelectrotonus in some cases becomes less marked and changes into relative anodic enhancement.

Later these phenomena were studied more fully in experiments on nerve trunks (Perna, 1912; Erlanger and Blair, 1931;
Chweitzer, 1937; Eichler, 1933; Parrack, 1940; Medvedenskii, 1940;
Khodorov, 1950) and on skeletal muscle (Perna, 1912). Further confirmation was obtained by investigation of the electronic changes
in the thresholds of excitability of single nerve fibers (Tasaki et.
al., 1948; Niedergerke, 1953) and of skeletal muscle fibers (Serkov, 1957).

These investigations showed that a "secondary" electrotonic increase in the thresholds beneath the cathode and a decrease beneath the anode arise within a few milliseconds after application
of the polarizing current; the worse the state of the preparation
studied and the stronger the steady current applied relative to the
threshold, the faster these changes take place.

Later work using a microelectrode technique showed (Frank
and Fuortes, 1956; Coombs, Curtis, and Eccles, 1959) that a sim
ilar rule applies to the nerve cells of the cat's spinal cord.

A number of suggestions have been made to explain the nature
of these electrotonic changes in excitability. Some workers (see
Ebbecke, 1933) have tried to link them with a change in permea
bility of the excitable membrane in the region of the cathode and
anode. Others claim that the electrotonic changes in excitability
are caused by a change in metabolism of the polarized tissue (Muzheev, Sviderskaya, and Shitova, 1937; Babskii, 1941; Vasilevskii,
1946).

A fundamentally different approach to the understanding of
these phenomena was suggested by Werigo as long ago as in 1888.
Werigo concluded on indirect evidence that Pflüger's electrotonic
changes in excitability are due to physical interaction between the
"electrotonic branches" of the polarizing and stimulating currents
in the nerve: to their summation when thresholds are measured

in the region of the cathode and to subtraction of one from the other when thresholds are measured in the region of the anode.

This physical summation, according to Werigo's theory, masks the development of a physiological process in the nerve leading to a progressive lowering of excitability beneath the cathode and to its increase beneath the anode. It is only if catelectrotonus persists for a very long time that the excitability falls to such a degree that it can no longer be masked by summation of the currents: in this period the thresholds of stimulation become higher than the initial value, and "overt" cathodic depression develops.

Werigo's ideas were not accepted by his contemporaries and were quickly forgotten (as was first pointed out by Arshavskii and Kurmaev, 1936). Experimental confirmation of Werigo's views on the origin of Pflüger's electrotonic changes in excitability was obtained by the investigations of Khodorov (1950) and of a group of Nasonov's collaborators: B. P. Ushakov, M. S. Averbakh, I. P. Suzdal'skaya, V. P. Troshina, and T. N. Cherepanova (Ushakov et al., 1953; see also Nasonov, 1959).

However, these investigations were conducted on whole nerve trunks, covered by a sheath and containing thousands of nerve fibers. The final decision had therefore to await direct experiments which had of necessity to be carried out on single excitable structures with the direct recording of the changes in membrane potential evoked by the polarizing and testing currents.

Such experiments were successfully performed on single nodes of Ranvier of isolated frog's nerve fibers and they completely confirmed the validity of the basic principles of Werigo's theory (Khodorov, 1962; Khodorov and Belyaev, 1963a,b; Khodorov, 1965b).

Recordings of the changes in the membrane potential of a single node of Ranvier, initially during rheobase stimulation before application of a steady current (test a), and later during application of a threshold testing stimulus a short time interval after the beginning of subthreshold depolarization (test b), and during hyperpolarization (test c) of the membrane, are illustrated in Fig. 1. It will be clear that the values of the threshold potential V_t measured during the first milliseconds of action of the current were lowered under the cathode and raised under the anode by comparison with their initial value. The reasons for these initial

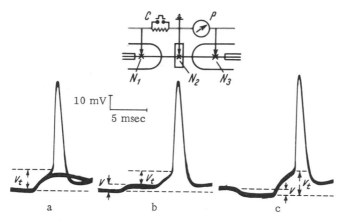

Fig. 1. Electrotonic changes in excitability of a medullated nerve fiber. The diagram above shows the arrangement of the nerve fiber on the electrodes. N_1 and N_3 are nodes of Ranvier inactivated with 0.3% procaine hydrochloride; N_2 is a node in Ringer's solutions; C is the circuit for stimulating and polarizing node N_2; P is the circuit for recording the potentials. a) Response of single node of Ranvier to pulse of steady current of subthreshold and threshold strength, b) and c) algebraic summation of changes in potential evoked by polarizing (V) and testing (ΔV) currents when excitability of the node is measured under the cathode (b) and under the anode (c). V_t is the threshold potential. The values of V are reduced to some degree by the shunting effect of the thin layer of liquid in the segment between the nodes (Khodorov, 1965b).

(Pflüger's) electrotonic changes in thresholds are made clear by analysis of these records.

A steady cathodal current induces subthreshold depolarization of the membrane and shifts the resting potential E_r toward the critical level E_c by an amount V. The testing stimulus simply completes this process by inducing extra depolarization of the membrane V_t. To attain threshold excitation of the node under the cathode, it is therefore sufficient to use a weaker stimulus than before the current began to act. The action potential in a single node during catelectrotonus appears as soon as the combined shift V + V_t reaches the critical level (Fig. 1b).

The resting potential E_r under the anode moves away from the critical level E_c by an amount V, and to reach the critical level a stronger testing stimulus must be used. The action potential appears as soon as the combined change in E under the anode $V_t - V$ becomes critical (Fig. 1c).

The first microphysiological investigations on isolated nerve fibers (Tasaki, 1956), on skeletal muscle fibers (Jenerick and Gerard, 1953), and on motoneutrons of the cat's spinal cord (Coombs, Eccles, and Fatt, 1955) created the impression that the critical potential is a stable value, remaining virtually unchanged during depolarization and hyperpolarization of the membrane.

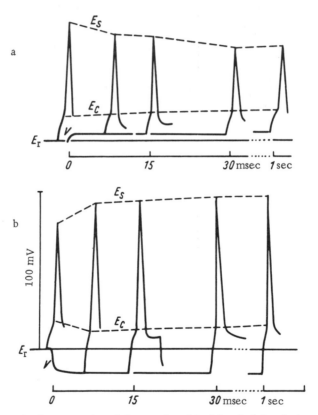

Fig. 2. Dynamics of changes in critical level of depolarization (E_C) and level of peak of action potential (E_S) of a node of Ranvier during prolonged depolarization (a) and hyperpolarization (b) of the membrane by a steady current. E_r is the initial resting potential level; V the change in potential induced by application of the current. The values of V are not corrected (see last sentence of caption of Fig. 1).

It was subsequently shown that this impression is valid only for the first few milliseconds (1-10 msec) of action of the steady current. During longer changes of potential the value of E_C under the cathode begins to rise, while that under the anode at first falls, but later (this is not observed in every case) it also increases (Fig. 2a,b).

The rate of the changes in E_c increases with an increase in the strength of the applied steady current, and for any given strength it depends on the temperature and functional state of the specimen.

For example, lowering the temperature sharply reduces the rate of the cat- and anelectrotonic changes in E_c ($Q_{10} \sim 3$). The initial period during which E_c stays at its initial level in some cases may extend over tens of milliseconds.

Mechanical injury to the node (for example, during dissection), an increase in the concentration of K^+ ions in the solution, or preliminary prolonged subthreshold depolarization of the membrane lead to a sharp increase in both the rate and degree of increase of E_c under the cathode and of its decrease under the anode.

These changes in E_c during electrotonus give rise to changes in the thresholds V_t which, as we know, are determined by the difference between the initial and critical values of the membrane potential.

Since the thresholds V_t are indices of the excitability of the fiber, we can conclude that "secondary" (accommodational, see p. 184) changes in excitability during cat- and anelectrotonus are the direct result of changes in the critical potential E_c: away from the resting potential level under the cathode and toward E_r under the anode.*

Experiments on single nerve fibers (Frankenhaeuser, 1952; Diecke, 1954; Müller, 1958), on skeletal muscle fibers (Jenerick and Gerard, 1953), on fibers of the conducting tissue of the myocardium (Weidman, 1955), and on motoneurons of the toad's spinal cord (Araki, 1960) further showed that, besides changes in excitability, during subthreshold depolarization of the membrane there

---

*For the relationships between V_t and the threshold strength of the current I_t, see Chapter VI.

is also a definite decrease in the amplitude and the steepness of rise of the action potentials.

During prolonged or strong catelectrotonus the action potentials disappear completely: instead of action potentials, only local responses arise to the testing stimuli.

The second new fact yielded by the experiments on single nodes of Ranvier was the discovery of the secondary anodic increase in the critical level of depolarization. During prolonged hyperpolarization of the membrane, after the initial fall in E_c, in some cases it begins to rise again (Fig. 2b), and the stronger the anodic current applied, the greater the rise in E_c (Fig. 3). The amplitude of the action potential and the steepness of its rise still remain increased during this anodic depression by comparison with their initial values (Khodorov, 1962, 1965a,b; Khodorov and Belyaev, 1966a,b).

The increase in E_c during hyperpolarization was also shown to be facilitated by an increase in the concentration of divalent cations in the medium: elevation of the concentration of Ca^{++} ions (Fig. 3) or addition of Ni^{++} or Cd^{++} ions to the solution (Khodorov, 1965b; Khodorov and Belyaev, 1966a,b).

The opposite effect was obtained by reducing the Ca^{++} ion concentration in the medium. In this case the anelectrotonic decrease of the critical level E_c becomes very much greater, and in some cases it reaches the initial level E_r even if the anodic current is comparatively weak. Merely to disconnect the hyperpolarizing current is now sufficient to allow the potential, which recovers its initial value rapidly, to reach E_c and to cause the generation of an action potential. This also explains the fact that the generation of anodic off-excitation takes place much more easily in a medium with a reduced Ca^{++} ion concentration (Frankenhaeuser and Widen, 1956).

The phenomena of cat- and anelectrotonus were rightly regarded by the pioneers in electrophysiology as a prototype, or model, of the processes of excitation and inhibition developing during natural activity of the nervous system (Werigo, 1901a,b; Wedensky, 1922; Ukhtomskii, 1950). Investigations have in fact shown that phenomena such as post-activity enhancement of excitability, synaptic facilitation, the increase in excitability of a neuron un-

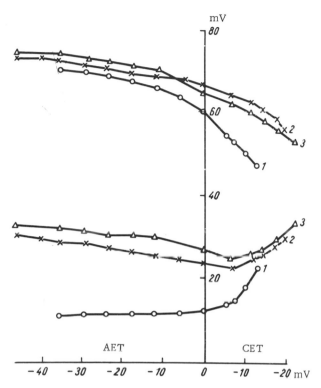

Fig. 3. Critical level of depolarization (bottom curves) and
amplitudes of action potential of a node of Ranvier (top
curves) plotted against catelectrotonic (CET) and anelectrotonic
(AET) changes in resting potential. Abscissa: changes in potential (V) in mV; ordinate: critical potential and spike amplitude measured from initial level of resting potential (in
mV). Values of potential not corrected. 1) Node N_2 in
Ringer's solution with normal Ca^{++} ion concentration (1.8
mM); 2) Ca^{++} concentration increased to 27 mM; 3) Ca^{++}
concentration increased to 36 mM (Khodorov and Belyaev,
1966b).

der the influence of its bombardment by subthreshold afferent impulses arising from different sources, and other similar phenomena due to subthreshold depolarization of the membrane are based
on the same mechanism, in principle, as this catelectrotonic increase
in excitability. On the other hand, it has been shown that cathodic
depression plays an important role in the mechanism of certain

types of peripheral and central inhibition based on excessively strong or prolonged depolarization of the membrane: the Wedensky "pessimum," presynaptic inhibition, inhibition in the cells of the hippocampus and cerebral cortex, and so on.

Let us now examine some of these phenomena in more detail.

Effect of Afferent Stimulation

Haapen, Kolmodin, and Scoglund (1958) compared the changes in the resting potential and critical potential of interneurons of the cat's spinal cord during their natural bombardment by afferent impulses.

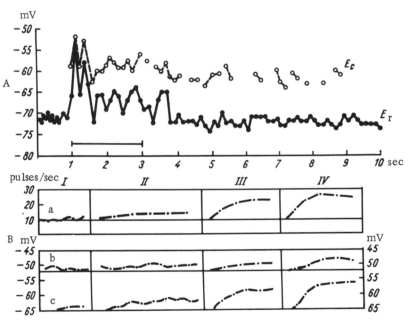

Fig. 4. Effect of afferent stimulation on excitability of interneuron of the cat's spinal cord. A: Graphical representation of variations in resting potential (E_r) and critical potential (E_c) of interneuron as the result of application of pad pressure (horizontal line). Single shock stimulation of afferent nerve starts at point marked 1st, second, and continues throughout the figure. Abscissa: time in msec; ordinate: inner membrane potential. Filled circles without corresponding open circles indicate absence of responses to stimulation (Haapen et al., 1958); B: diagram showing change in frequency (a), critical potential (b), and slow depolarization of membrane (c) of motoneuron during four successive periods of its proprioceptive activation at increasing strength (I-IV) (Kolmodin and Scoglund, 1958).

In these experiments depolarization of the membrane was accompanied by an increase in the level of E_c, while hyperpolarization led to its decrease. However, E_c always changed by a smaller amount than E_r, so that the threshold of depolarization V_t was lowered in the first case (excitability was increased), and was raised in the second (excitability was lowered) (Fig. 4A).

Kolmodin and Scoglund (1958) also investigated the dependence of the critical potential on the firing rate of the interneuron in response to afferent stimulation of different strengths. As Fig. 4B shows, with an increase in the intensity of stimulation (I-IV) the firing rate (a) and the critical potential (b) both increase.

However, even though the firing rate was more than doubled (Fig. 4B, IV), the critical level was not changed by more than 4 mV.

Wedensky's "Pessimal" Inhibition

Araki and Otani (1959) used intracellular microelectrodes to study the response of motoneurons of the toad's spinal cord to orthodromic stimuli of different frequencies. Their experiments showed that if the interval between orthodromic stimuli was reduced from 22.6 to 18.0-11.3 msec, a Wedensky's "pessimum" develops in the cells. Definite summation of the postsynaptic potentials is found under these circumstances; it leads to persistent depolarization of the membrane and, as a result, to elevation of the critical potential.

The observations of Araki and Otani (1959) are in agreement with the results of an investigation by Shamarina (1961), who

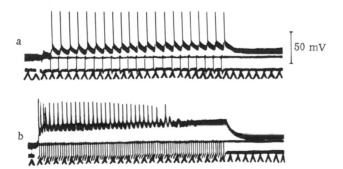

Fig. 5. Changes in critical potential of a skeletal muscle fiber during stimulation of the motor nerve at optimal (a) and "pessimal" (b) frequencies. Time marker, 20 msec (Shamarina, 1961).

studied the mechanism of "pessimal" inhibition on a frog nerve-muscle preparation (Fig. 5).

By taking intracellular recordings exactly from the synaptic region of the muscle fiber, Shamarina found that if the nerve is stimulated at frequencies of 50–100/sec, the amplitude of the action potentials falls away while, at the same time, the depolarization of the postsynaptic membrane rises gradually.

These results support the view that the Wedensky "pessimum" is in fact related to the phenomenon of Werigo's cathodic depression. Depolarization of the postsynaptic membrane at the "pessimum" gives rise to a twofold effect: elevation of the critical potential and depression of the ability of the postsynaptic membrane to generate action potentials.

Relative Refractoriness

In experiments on single nodes of Ranvier of frog's nerve fibers, Tasaki (1956), Dodge (1961), and other workers have found that there is a considerable increase in the thresholds of depolarization V_t in the phase of relative refractoriness, even after the membrane potential and membrane resistance have returned to their initial values.

It can accordingly be concluded that the increase in V_t taking place in the medullated nerve fiber is due to an increase in the critical potential E_c.

In the giant axon of the squid the phase of relative refractoriness coincides in time with after-hyperpolarization of the membrane, so that the increase in V_t in this phase is due not only to the increase in E_c, but also to a decrease in E_r (Hodgkin and Huxley, 1952d; Timin and Khodorov, 1971).

Effects of Potassium Ions

The changes in the resting potential and critical potential of the nerve fiber with an increase in the concentration of K^+ ions in the medium are very similar to those observed in catelectrotonus.

The results of one of a series of experiments carried out by Belyaev on a single node of Ranvier from a frog nerve fiber (Rana temporaria) are illustrated in Fig. 6.

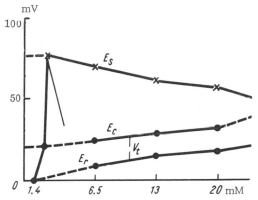

Fig. 6. Changes in resting potential (E_r), critical potential (E_C), and spike potential (E_s) of Ranvier node with an increase in KCl concentration in solution. Abscissa:, KCl concentration in mM; ordinate: changes in E_r, E_C, and E_s in mV (values not corrected; see last sentence of caption of Fig. 1).

Increasing the KCl concentration in the solution from 1.4 to 20 mM not only depolarized the membrane, but also increased E_c. However, the changes in F_c in this case were rather less than the changes in E_r, and for this reason the thresholds V_t were reduced by the action of the K^+ ions. The steepness of rise of the action potential and its amplitude also were reduced. In a medium containing 27 mM KCl the node completely lost its ability to generate action potentials. Only a local response took place to stimulation. Rinsing the node completely restored the initial level of the resting potential and the initial amplitude of the action potential, and only F_c remained slightly increased by comparison with its initial value.

A similar result was also obtained in other experiments. In some cases, however, the increase in E_c under the influence of an excess of K^+ ions was more marked than the change in E_r, and in that case V_t remained increased from the very beginning.

Changes in V_t Resulting from a
Primary Shift of E_c

Effects of Calcium Ions

It has been known for a long time that Ca^{++} ions have a specific effect on the excitability of nerve and muscle fibers. An increase in $[Ca]_0$ increases the thresholds, while a decrease in $[Ca]_0$, on the contrary, causes a decrease in the thresholds and in some cases the appearance of "spontaneous" activity in these structures (Brink, 1954).

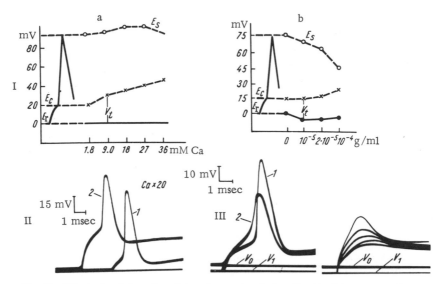

Fig. 7. Effect of calcium ions (a) and procaine hydrochloride (b) on generation of action potentials in a node of Ranvier. a: I) abscissa: $CaCl_2$ concentration in mM; ordinate: changes in E_r, E_c, and E_s in mV; II: 1) response of node in Ringer's solution containing 1.8 mM $CaCl_2$; II: 2) the same after an increase in the $CaCl_2$ concentration to 36 mM. Stimulation of node by a steady current (Khodorov, 1965a); b: I) abscissa: concentration of procaine in g/ml; ordinate: changes in E_r, E_c, and E_s in mV; II: 1) response of node in Ringer's solution; II: 2) on addition of procaine hydrochloride to solution to a concentration of 2 $\times$ 10^{-5} g/ml; III) gradual responses to 10^{-4} g/ml procaine. V_0 is initial level of resting potential, V_1 change in resting potential toward hyperpolarization under the influence of procaine. Values of membrane potential are not corrected (Khodorov and Belyaev, 1965b).

These changes in excitability under the influence of Ca^{++} are mainly determined by changes in the level of E_c.

An increase in the concentration of Ca^{++} ions (by 5–20 times) in the Ringer's solution does not alter the resting potential of nerve fibers (Frankenhaeuser and Hodgkin, 1957; Frankenhaeuser, 1957b; Khodorov, 1962, 1965a,b), while at the same time it considerably raises the critical level of depolarization (Fig. 7a). As a result of this, V_t rises and excitability falls. The amplitude of the action potentials, on the other hand, usually is increased (Ichioka, 1957; Khodorov, 1965a,b). A decrease in the concentration of Ca^{++} ions has the opposite effect: the critical level of depolarization and the amplitude of the action potentials fall, while the excitability is increased. If, in a medium with a reduced calcium concentration, lowering of E_c is accompanied by depolarization of the membrane (an increase in E_r), as happens in muscle (Shapovalov, 1960) and some nerve fibers (Stämpfli and Nishie, 1956; Schmidt, 1964), spike discharges appear in an independent rhythm; as a rule they arise on the crest of the slow depolarization wave (Shapovalov, 1960).

Calcium ions have a similar effect on E_c on skeletal muscle fibers. In this case, however, besides the increase in E_c under the influence of Ca^{++} ions, hyperpolarization of the membrane also takes place (Vornovitskii, 1966), and the increase in V_t in skeletal muscle fibers is thus due not only to changes in E_c, but also to definite changes in E_r.

The ability to increase the critical level of depolarization of nerve and muscle fibers despite the unchanged resting potential is not restricted to Ca^{++} ions, but it is also possessed by other divalent cations, including Ni^{++}, Cd^{++}, and Co^{++}. However, as well as increasing V_t, these other ions also give rise to specific changes in the shape of the action potential (see p. 120).

Effects of Local and General Anesthetics

Local anesthetics (procaine, cocaine, etc.), like Ca^{++} ions, have little effect on the initial level of the resting potential but considerably increase the critical potential, as a result of which the threshold depolarization ($V_t = E_c - E_r$) is raised. However, unlike Ca^{++} ions, local anesthetics have a strongly inhibitory action on the generation of action potentials in excitable structures (Fig. 7b).

For instance, single nodes of Ranvier from frog nerve fibers instantaneously lose their ability to generate action potentials when exposed to the action of 10^{-4} g/ml procaine hydrochloride; only local responses arise in the node to testing stimuli, and these responses increase in size with an increase in the strength of the stimulus (Fig. 7b).

In lower concentrations of procaine (10^{-5} g/ml) single nodes can still generate action potentials, but the amplitude and steepness of their ascending phase are considerably reduced. In some cases the responses of the Ranvier nodes to stimulation become graded in character: weak stimuli elicit local responses, stronger stimuli evoke spikes, the amplitude of which does not obey the "all or nothing" rule (Khodorov and Belyaev, 1965b).

Curtis and Phillis (1960) used a technique of ionophoresis to apply procaine hydrochloride via an extracellular microelectrode to the surface of motoneurons, interneurons, and Renshaw cells of the spinal cord, while simultaneously, through another (intracellular) microelectrode, they recorded their responses to direct (electrical, acetylcholine, glutamic acid) and indirect (orthodromic and antidromic) stimulation. These experiments showed that although procaine does not alter the resting potential or resistance of the membrane of nerve cells, it increases the thresholds of their depolarization to direct, orthodromic, and antidromic stimulation and sharply reduces the amplitude of the action potentials. Similar effects on the relationship between E_c and E_r are produced by barbiturates (Blaustein, 1968), tetrodotoxin (see p. 116), and certain other agents.

The results described above thus show that the relationship between the values E_r and E_c is not constant and invariable. Depending on the direction and extent of the change in the level of the resting and critical potentials induced by a particular external or internal environmental factor, the threshold potential V_t may decrease, increase, or remain unchanged.

However, both the resting potential and the critical potential are dependent on the ionic permeability of the excitable membrane, and it is therefore natural to try to find a direct link between excitability and the various constants of the ionic permeability of the membrane.

This problem will be examined in the subsequent chapters.

Chapter II

The Resting Potential

The role of the cell membranes in the genesis of bioelectrical phenomena was first postulated by Ostwald (1890). On the basis of the results of his experiments with sedimentation membranes Ostwald postulated that "semipermeable membranes are the place where a sudden change of potential arises ..." and he later went on to say that "not only currents in muscles and nerves, but also the puzzling actions of electric fishes can also be explained by... the properties of semipermeable membranes."

Ostwald's idea was developed by Bernstein, a pupil of DuBois Roymond.

In the hypothesis put forward by Bernstein (1902, 1912) the membrane of nerve and muscle fibers at rest is selectively permeable only to K^+ ions, the concentration of which in the cytoplasm is much higher than in the external medium. The resting potential can therefore be regarded as a potassium diffusion potential, the magnitude of which is determined by the formula

$$E_M = E_K = \frac{RT}{F} \ln \frac{[K]_i}{[K]_0} = 58 \log \frac{[K]_i}{[K]_0} \quad (\text{at } 18° \text{C}) \qquad (2)$$

Here E_M is the transmembrane potential difference, E_K the potassium equilibrium potential, i.e., the potential at which the diffusion flow of K^+ ions with the concentration gradient becomes equal to the flow of K^+ ions in the opposite direction due to the electric field (produced by the diffusion of these ions); R is Clapeyron's constant; T the absolute temperature; F Faraday's constant; $[K]_i$

and $[K]_0$ the activities or concentrations of K^+ ions inside and outside the cell respectively. Bernstein's hypothesis explained many
of the facts known to physiologists at that time quite satisfactorily,
including the dependence of the resting potential on the K^+ ion concentration in the medium and on the absolute temperature. However, it could not be verified by direct experiment for several decades because no method was then available for measuring the absolute magnitude of the resting potential directly or for determining the permeability of a membrane to different ions.

For many years the membrane theory was severely criticized
by the supporters of the phase theory of biopotentials (see: Nasonov and Aleksandrov, 1944; Nasonov, 1949, 1959; Troshin, 1956;
Ling, 1960; Kurella, 1960). It is impossible here to examine fully
the various stages of this prolonged discussion (see the surveys
by Liberman and Chailakhyan, 1963a,b). Only the development and
introduction of intracellular microelectrode techniques, followed
by techniques of intracellular perfusion of giant axons (see below),
convinced most investigators that Bernstein's membrane theory is
basically correct although it needs substantial additions and modifications.

The first discovery which was made was that the cell membrane at rest is permeable not only to K^+ ions, but also to Cl^- ions
(Fenn, 1936; Fenn and Haeg, 1942; Boyle and Conway, 1941) and to
Na^+ ions (Heppel, 1939; Manery and Ball, 1941; Ussing, 1947).

In most nerve fibers the chloride permeability (P_{Cl}) is much
lower than the potassium permeability (P_K). However, in skeletal
muscles, at the normal resting potential, the value of P_{Cl} is about
twice that of P_K (Hodgkin and Horowicz, 1959a,b, 1960). Permeability of the membrane of nerve fibers to Na^+ ions (P_{Na}) at
rest is about a hundred times lower than the permeability for K^+
ions.

It was shown later that in most excitable structures there are
definite differences between the value of the resting potential
measured by means of intracellular electrodes and the corresponding value of the potassium equilibrium potential calculated
by Nernst's formula.

For instance, in the giant axons of the squid, as Steinbach and
Spiegelman (1943) found, E_K is -75 mV, while the resting potential varies from -50 mV (in isolated axons removed from the

animal) to -70 mV (in intact fibers connected to their blood supply). In skeletal muscle fibers $E_K = -101.5$ mV, and the resting potential is -80 or -90 mV (Conway, 1957; Adrian, 1956, 1960; Hodgkin and Horowicz, 1959a).

In medullated nerve fibers the K^+ ion concentration inside the axon is taken to be 120 mM (it has not yet been measured exactly). If $K_0 = 2.5$ mM, then E_K for these fibers is -97.5 mV (Frankenhaeuser, 1962a). The measured value of the resting potential, however, is -70 mV (Huxley and Stämpfli, 1951a; Frankenhaeuser, 1960) or -75 mV (Dodge, 1963).

Finally, it has been shown that the linear relationships between $\ln([K]_i/[K]_0)$ and the resting potential (E_M) predicted by Bernstein's hypothesis in fact exist only if the K^+ ion concentration in the medium exceeds 10 mM. At lower K^+ concentrations the curve of E_M versus $[K]_0$ diverges from the theoretical. These results were obtained on giant axons of the squid (Curtis and Cole, 1942; Fig. 8a), skeletal muscle fibers (Ling and Gerard, 1950; Adrian, 1956), and single nodes of Ranvier from frog's nerve fibers (Huxley and Stämpfli, 1951b; Fig. 8b).

To explain all these divergences, Hodgkin and Katz (1949) suggested that the resting potential of excitable structures is determined not just by the potassium, but also by the sodium and chloride equilibrium potentials, and that the contribution of each of these potentials to the resting potential is determined by the ratios between the permeabilities of the excitable membrane for these ions.

When P_K is much greater than P_{Na} and P_{Cl}, the resting potential is close to the potassium equilibrium potential. If, however, P_{Na} of the membrane is increased and, consequently, the inflow of Na^+ ions is intensified, the transmembrane potential difference is lower than E_K.

To develop the quantitative aspect of this hypothesis, Hodgkin and Katz (1949) made use of the "constant field" theory (Goldman, 1943).

The main points of this theory are: 1) ions move in a membrane under the influence of diffusion and the electric field in a manner which is essentially similar to that in free solution; 2) the intensity of the electric field may be regarded as constant throughout the membrane, i.e., the fall of potential across the thickness

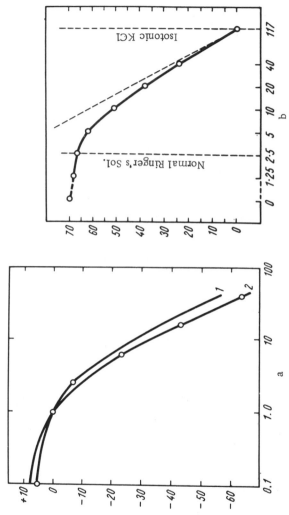

Fig. 8. Resting potential of the squid giant axon (a) and medullated nerve fiber (b) as a function of the K+ ion concentration in solution. a) Abscissa: K+ ion concentration in multiples of original concentration in sea water (13 mM), logarithmic scale; ordinate: changes in transmembrane potential difference relative to original, taken as zero. Circles indicate experimental results; curves 1 and 2 calculated by the "constant field" equation (see text) (Hodgkin and Katz, 1949). b) Abscissa: K+ ion concentration in mM, logarithmic scale; ordinate: transmembrane potential difference (E_M) in mV. Gradient of curve E_M versus log $[K]_0$ approximates to the theoretical (broken line) at high values of $[K]_0$ (Huxley and Stämpfli, 1951b).

of the membrane is linear; 3) the concentrations of ions at the edges of the membrane are directly proportional to those in the aqueous solutions bounding the membrane; 4) the membrane is homogeneous, i.e., it has the same dielectric constant throughout.

It follows from point 1 that the density of the ionic current (I) of a given ion i through a membrane is given by

$$I_i = -z_i u_i RT \frac{dC_i}{dx} - E z_i^2 u_i C_i \frac{dV}{dx},$$ (3)

where z_i is the valence of the ion i, u_i its mobility, C_i its concentration, and V the potential at point x (the distance across the thickness of the membrane from its outer border, from the coordinate x = 0).

This means that the density of the ionic current is determined by the difference between the diffusion flow of the ion i along the concentration gradient ($-z_i u_i RT dC_i/dx$) and the flow of this ion in the opposite direction due to the electric field arising from diffusion ($E z_i^2 u_i C_i dV/dx$).

If the electric field on the membrane is constant: dV/dx = const = V/a, where a is the thickness of the membrane. In this case the constant of permeability of the membrane for ion 1 (P_i) is determined by its relative mobility and by its solubility in the material of the membrane:

$$P_i = u_i \beta_i \frac{RT}{aF},$$ (3a)

where β_i is the partition coefficient of ion i between the aqueous solution and the membrane.

By means of a series of transformations Hodgkin and Katz obtained equations expressing the relationship between the densities of the potassium (I_K), sodium (I_{Na}), and chloride (I_{Cl}) currents, on the one hand, and the membrane potential and ionic permeability (P_K, P_{Na}, and P_{Cl}) on the other hand:

$$I_K = P_K \frac{EF^2}{RT} \frac{[K]_0 - [K]_i \exp\{EF/RT\}}{1 - \exp\{EF/RT\}}$$ (4)

$$I_{Na} = P_{Na} \frac{EF^2}{RT} \frac{[Na]_0 - [Na]_i \exp\{EF/RT\}}{1 - \exp\{EF/RT\}} \tag{5}$$

$$I_{Cl} = P_{Cl} \frac{EF^2}{RT} \frac{[Cl]_i - [Cl]_0 \exp\{EF/RT\}}{1 - \exp\{EF/RT\}} \tag{6}$$

In a resting state, in a homogeneous part of the membrane, the resultant ionic current $(I_K + I_{Na} + I_{Cl})$ is equal to zero, and the potential inside the membrane is given by

$$E = \frac{RT}{F} \ln \frac{P_K[K]_0 + P_{Na}[Na]_0 + P_{Cl}[Cl]_i}{P_K[K]_i + P_{Na}[Na]_i + P_{Cl}[Cl]_0}. \tag{7*}$$

"There are many reasons," Hodgkin and Katz state, "for supposing that this equation is no more than a rough approximation, and it clearly cannot give exact results if ions enter into chemical combination with carrier molecules in the membrane or if appreciable quantities of current are transported by ions other than K, Na, or Cl. On the other hand, the equation has two important advantages. In the first place it is extremely simple, and in the second it reduces to the thermodynamically correct forms when any one permeability constant is made large compared to the others." (Hodgkin and Katz, 1949, p. 66.)

Hodgkin and his followers later repeatedly emphasized that to deduce this formula it was necessary to accept only the first principle of Goldman's theory, i.e., that ions move in the membrane under the influence of diffusion and the electric field just as they do in free solution. Constancy of the electric field of the membrane is not necessary in this case.

By substituting the concentrations of K^+, Na^+, and Cl^- ions in the cytoplasm known from chemical measurements (Steinbach and Spiegelman, 1943) in the equation, Hodgkin and Katz (1949) found that in the resting giant nerve fiber, kept in a solution of artificial seawater, the ratio between the permeability constants is

$$P_K : P_{Na} : P_{Cl} = 1 : 0.04 : 0.45.$$

These ratios still held good when the value of $[K]_0$ was increased or decreased within the range from 1 to 15 mM (Fig. 8a, curve 1).

---

*In Eqs. (5)-(7) the signs of the current and potential are changed in accordance with the currently accepted convention.

With an increase in concentration from 15 to 150 mM, however, when strong depolarization of the membrane developed (Fig.
8a, curve 2), the ratio between the permeability constants was
changed so that P_K became even higher relative to P_{Na} and P_{Cl}:

$$P_K : P_{Na} : P_{Cl} = 1 : 0.025 : 0.3.$$

The fact that the experimental curve of E versus log $[K]_0$ de-
viated from the theoretical curve calculated by Nernst's formula
in the region of low values of $[K]_0$, discovered by Curtis and Cole
(1945) and by other investigators, was thus explained. The reason for this divergence, according to the views examined above,
is that the ratios between P_K, P_{Na}, and P_{Cl} are all dependent to a
definite degree on the level of the membrane potential which, in
turn, depends on the concentration of these ions in the medium.

Still more precise evidence of the dependence of the values
of P_K, P_{Na}, and P_{Cl} on the level of the membrane potential was
obtained later.

To study the effect of the ionic composition of the medium on
the resting potential and action potential of the medullated nerve
fiber, Stämpfli (1958, 1959) developed a technique of external perfusion of a node of Ranvier. This technique made it possible to
change the salt solutions bathing the node in the course of about
0.2 sec, so that it was possible not only to record the final level
to which the resting potential shifted under the influence of an excess of K^+ ions in the solution, but also the rate at which this
change developed.

The experiments revealed substantial differences between the
sensitivity of different nerve fibers, and sometimes of the same
nerve fiber in different states, to a moderate increase in the KCl
concentration in the solution.

If the nerve fiber had been removed rapidly and with every
possible precaution, an increase in $[K]_0$ to 20 mM or, in certain
cases, to 40 mM produced only slight, slowly increasing depolarization, and not until a certain critical K^+ concentration was reached
(usually > 20 mM) did the depolarization take place suddenly to
reach the level of the potassium equilibrium potential.

This low sensitivity of the node membrane to K^+ ions usually
disappeared after the repeated action of KCl in high concentrations

on the fiber or after depolarizing electrical stimulation of the membrane.

If the fiber was less carefully dissected, the reaction to an increase in $[K]_0$ was steplike in character from the very beginning: depolarization of the membrane increased rapidly and the resting potential reached a level close to the potassium equilibrium potential calculated from Nernst's formula. The high sensitivity of these fibers to an excess of K^+ ions could be suppressed by preliminary hyperpolarization of the membrane. In this case, application of KCl to the node, instead of producing a sudden shift in the potential, at first caused only a slight and slowly increasing depolarization; not until a few seconds later did the potential E_r suddenly begin to rise rapidly, although it did not reach the previous maximum of depolarization.

Stämpfli (1959) considered that the reason for the differences in sensitivity of nerve fibers to potassium ions is their unequal potassium permeability. The intact fiber has a low P_K, and moderate changes in $[K]_0$ thus have little effect on the resting potential. If, however, the fiber is damaged during dissection, it has a high initial P_K, and the membrane of such a fiber thus behaves like a potassium electrode. Hyperpolarization of the membrane reduces the previously raised P_K and thus restores the original low sensitivity of the membrane to an excess of K^+ ions. Stämpfli's (1959) results for the effect of anelectrotonus on the membrane sensitivity of the node of Ranvier to K^+ ions were subsequently confirmed by Ulbricht (1963), who made improvements to the system for perfusing the node. In Ulbricht's experiments a complete change of the fluid bathing the node took 10 msec, so that the process of depolarization of the membrane under the influence of 40 mM KCl was complete within several tenths of a second. By recording the dynamics of this process, Ulbricht clearly demonstrated that the potassium depolarization of the membrane is retarded or even suppressed during anelectrotonus (Fig. 9). Simultaneously with the abolition of depolarization, the anodic current also restores the resistance of the membrane.

A similar, but somewhat less marked, effect in potassium depolarization is produced by Ca^{++} ions: they retard its development and slightly lower the level which it reaches during the prolonged action of an excess of KCl (see also Schmidt, 1964).

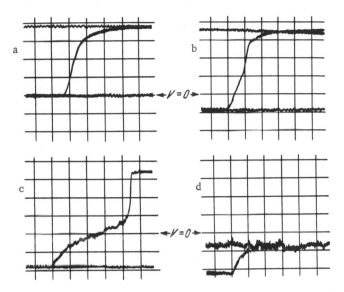

Fig. 9. Abolition of depolarization of the membrane of a node of
Ranvier induced by 42.5 mM KCl, by a hyperpolarizing current.
Each division along the abscissa represents a 100 msec (a-c); 200
msec (d); along the ordinate each division represents a change in
potential of 8.5 mV. Arrow indicates initial level of resting potential in Ringer's solution (V = 0). a) Before action of current,
b) during passage of weak hyperpolarizing current, c) moderately
strong current, d) during strong hyperpolarization. Ever-increasing
delay of the first phase of depolarization can be seen (b and c);
during the passage of a strong hyperpolarizing current the change
of potential does not even reach the level V = 0. "Agitation" of
the membrane can be seen during the initial phase of depolarization. Temperature 24°C (Ulbricht, 1963).

Results showing the effect of the quaternary ammonium base
tetraethylammonium on potassium depolarization of the membrane
are also directly relevant to the present issue. As investigations
by several workers have shown (p. 138), this compound specifically
blocks the potassium channels of the active membrane. This is
manifested, in particular, by the fact that addition of tetraethylammonium (up to 5-20 mM) to a solution containing an excess of
K^+ ions sharply reduces the potassium depolarization of the membrane of the Ranvier node, or even abolishes it if $[K]_0$ is low (Tasaki, 1959b). The membrane resistance, when lowered by the ac-

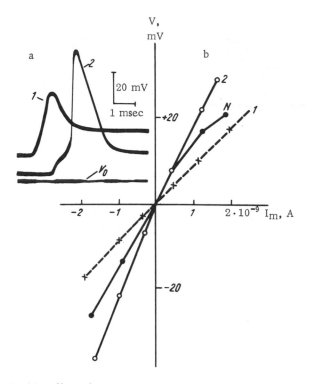

Fig. 10. Effect of tetraethylammonium chloride (TEA) on a single node of Ranvier in 30 mM KCl solution. a: V_0) initial resting potential, 1) response of node to threshold stimulation in 30 mM KCl solution, 2) the same, after addition of 20 mM TEA to the KCl solution; b) relationship between strength of current applied and change in membrane potential of Ranvier node. N: node in Ringer's solution of normal composition; 1) in 30 mM KCl solution; 2) in 30 mM KCl + 20 mM TEA solution. Abscissa: strength of depolarizing (positive values of I_m) and hyperpolarizing (negative values of I_m) currents. Ordinate: changes of potential in mV, measured 50 msec after applying current (Khodorov, 1965b).

tion of K^+ ions, rises as a result of this treatment, up to the same values as during strong hyperpolarization of the membrane (Khodorov, 1965b; Khodorov and Belyaev, 1965a; Fig. 10).

These results can easily be explained on the assumption that potassium depolarization of the membrane is a regenerative process.

An increase in $[K]_0$ initially, when P_K of the resting membrane is still low and, consequently the ratios P_K/P_{Na} and P_K/P_{Cl} are also relatively low, induces only weak depolarization of the membrane. This, however, increases P_K, and thus leads to an even greater decrease in the transmembrane potential difference. This course of events continues until depolarization reaches a level close to that of the potassium equilibrium potential.

Hyperpolarization of the membrane by an anodic current, an increase in $[Ca]_0$ and, in particular, tetraethylammonium all inhibit the increase in P_K, and the potassium depolarization is thus strongly retarded and weakened by the action of these agents. Conversely, mechanical injury to the membrane, its depolarization by a cathodic current, and a preliminary, prolonged exposure to a high $[K]_0$ increase P_K, and thereby accelerate the development of potassium depolarization.

New opportunities for studying the nature of the resting potential became available to physiologists after Hodgkin's (Baker and Shaw, 1961) and Tasaki's (Oikawa et al., 1961) collaborators successfully developed their technique of internal perfusion of giant nerve fibers with artificial salt solutions.

Baker, Hodgkin, and Shaw (1962) showed that if the cytoplasm is extruded from the squid giant axon and replaced by isotonic KCl or K_2SO_4 solution, a potential difference of the order of 50-60 mV, corresponding to the resting potential of the unperfused fiber, is established between the inside and outside of the membrane.

These experiments further showed that the presence of K^+ in the internal solution is an essential condition for the existence of the resting potential. If K^+ is removed from the perfusion fluid and replaced by Na^+, the potential difference between the internal contents of the fiber and the surrounding solution disappears completely. This disappearance of the resting potential, however, occurs only if the concentration of K^+ ions in the seawater bathing the outside of the axon is normal. If the axon is placed in isotonic KCl solution, and if NaCl is present inside the fiber, the sign of the resting potential is reversed: the inside of the membrane acquires a positive potential relative to the outside (+ 50 mV).

These facts directly confirmed the view that K^+ ions play the leading role in the genesis of the resting potential. At the same time, these experiments showed that the relationship between the resting potential and the internal K^+ ion concentration is complex in character.

For example, at $[K]_0 = 10$ mM, an increase in the K^+ ion concentration in the internal solution from zero to 250 mM (replacement of NaCl by KCl) causes a rapid decrease in the internal potential from 0 to -50 mV, but with a further increase in $[K]_i$ to 600 mM this potential falls by only 5-10 mV, instead of by the 35 mV expected for a potassium electrode.

This distinctive saturation phenomenon is very similar to what Curtis and Cole (1942) observed in their experiments with changes in the external K^+ ion concentration (Fig. 8a).

In their later investigations, Baker, Hodgkin, and Meves (1964) changed the internal K^+ ion concentration by replacing the KCl not by NaCl but by sucrose. In this case a decrease in KCl concentration from 600 to 300-100 mM often induced slight hyperpolarization of the membrane (by 10-15 mV), evidently through a decrease in the internal Cl^- ion concentration. At greater dilutions of KCl the hyperpolarization was replaced by depolarization of the membrane, and at 2 mM KCl the resting potential approximated to zero.

These workers attempted to determine the relative coefficients of potassium, sodium, and chloride permeability by comparing the mean results with values calculated from the constant field equation. The calculations showed that at $[KCl]_i$ values below 50 mM, the changes in the resting potential can be described on the assumption that

$$P_K : P_{Na} : P_{Cl} = 1 : 0.035 : 0.02.$$

If the internal KCl concentration was higher (from 50 to 600 mM), the relationship between E_r and $[K]_i$ was satisfied by a different ratio between the coefficients of permeability:

$$P_K : P_{Na} : P_C = 1 : 0.05 : 0.1.$$

This means that in the region of high values of [KCl], the membrane does not differentiate so well between K^+ and other ions as at low values of [KCl] and correspondingly high value of E_r.

An increase in $[K]_i$ to 695 mM reduced the measured value of E_r to -55.2 mV compared with the 103.5 mV calculated from Nernst's formula.

Using the same method, Adelman and Fok (1964) discovered that changes in the resting potential observed during variation of the K^+ ion concentration in the perfusion fluid from 5 to 456.6 mM (corresponding to a change in the activity of K^+ ions from 4.1 to 232.9 mM) can be satisfactorily described if the following ratio is assumed between the coefficients of ionic permeability:

$$P_K : P_{Na} : P_{Cl} = 1 : 0.06 : 0.25.$$

Investigations conducted on the giant axons of the squid, perfused internally with salt solutions of different composition, thus confirmed the views regarding the nature of the resting potential of nerve fibers previously suggested on the basis of experiments in which the external concentrations of K^+, Na^+, and Cl^- ions were changed.

So far we have spoken entirely of the role of passive ionic currents in the genesis of the resting potential. The question naturally arises: what is the role of active transport of Na^+ and K^+ ions through the membrane in this process?

In the beginning it was postulated that active ion transport has only an indirect effect on electrical activity of excitable structures: it creates and maintains the constancy of concentration gradients between the cytoplasm and the medium surrounding the cell. In this view the sodium—potassium pump has no direct effect on electrogenesis for in practice it is electrically neutral (Hodgkin and Keynes, 1955a).

Nowadays, however, this hypothesis has been revised. The work of Baker and Shaw (1965) on the giant nerve fibers of the squid and on nerve fibers of the crab have shown that the splitting of each high-energy phosphorus bond of adenosinetriphosphate is accompanied by the departure of 3 or 4 Na^+ ions from the cell. At the same time, it has been shown that the number of Na^+ ions "pumped" from the nerve fiber does not correspond exactly to the number of K^+ ions entering the cytoplasm. In squid giant axons, according to the observations of Caldwell et al. (1960), potassium exchange for sodium does not take place in the ratio 1:1, but in such a way that for each K^+ ion entering the axon 2 or 3 Na^+ ions leave the cell.

It follows from these results that the sodium—potassium pump is by no means electrically neutral, and under certain conditions it can generate a potential difference on the cell membrane.

In experiments on ganglia of the snail Helix pomatia, Kerkut and Thomas (1965) found that injection of Na^+ ions inside the neuron leads to the appearance of prolonged hyperpolarization, which was sensitive to variations of temperature, to removal of K^+ ions from the medium, and to the action of ouabain, a drug which specifically inhibits the operation of the sodium pump.

Experiments on various fibers from amphibians and warm-blooded animals have shown conclusively that the prolonged after-hyperpolarization of the membrane which develops after tetanic stimulation (Gasser, 1935) is connected with active transport of Na^+ ions which accumulate in the cytoplasm during activity (Connelly, 1959, 1962; Ritchie, 1961; Straub, 1962). This hyperpolarization is accompanied by increased oxygen demand (Connelly, 1959). Ouabain, in a concentration of 0.5 mM, completely suppresses post-tetanic hyperpolarization and reduces the excess oxygen demand by 90% (Connelly, 1962).

An electrogenic action of the sodium pump has also been found in skeletal muscle fibers enriched with Na^+ ions by preliminary and prolonged keeping in potassium-free Ringer's sulfate solution or at a low temperature (Mullins and Awad, 1965; Adrian and Slayman, 1966; Kernan, 1962, 1966).

However, it must be emphasized that in all the cases described above electrical manifestations of activity of the sodium pump were found only when the cytoplasm contained a much higher concentration of Na^+ ions than normally. If the Na^+ ion concentration in the nerve or muscle fibers was normal, the active transport of Na^+ ions was too small to induce any easily measurable fall of potential on the resistance of the membrane. This accounts for the fact that treatment with dinitrophenol, cyanide, or ouabain in concentrations high enough to suppress the sodium—potassium pump has no substantial effect on the resting potential of the excitable structures mentioned (Hodgkin and Keynes, 1955a; Adrian and Slayman, 1966; Baker, 1966).

The passive resistance of the membrane of these cells is about one order of magnitude higher than R_m of the membrane of the squid giant axon (Arshavskii, Berkinblit, and Khodorov, 1969, pp. 40–41), and active transport of sodium ions can thus make a substantial contribution to the magnitude of the resting potential.

According to Gorman and Marmor (1970), for instance, the electrogenic sodium pump accounts for approximately one-third of the total magnitude of the resting potential (~ -75 mV) at 17-20°C in the giant neurons of the marine mollusk <u>Anisodoris</u> <u>nobilis</u>. These workers came to this conclusion after a careful analysis of the effect of changes in temperature, $[K]_0$, $[Na]_0$, and ouabain on the resting potential. During exposure to all these factors the resting potential behaved as the sum of the "metabolic" and "ionic" components. In the case of suppression of the first component by ouabain or by cooling (to 4°C), changes in the ionic component were satisfactorily predictable by the constant field equation (7).[*]

---

[*] To judge from the latest results obtained by Casteels et al. (1971), the electrogenic Na pump can help to maintain the resting potential in smooth muscle cells (taenia coli of the guinea pig) also.

Chapter III

Passive and Subthreshold Active Changes in the Membrane Potential

Changes in the membrane potential produced by an electric current are usually subdivided into passive and active.

Passive, or electrotonic, changes in membrane potential are determined by the passive electrical characteristics of the membrane itself and of the cell (fiber) as a whole.

Passive changes in the membrane potential arise in response to the action of an electric current of any strength, shape, or direction on excitable structures. However, whereas passive changes of potential can be observed in what can be called a pure or virtually uncomplicated form during the passage of hyperpolarizing (anodic) and weak depolarizing (cathodal) currents, in response to near-threshold and above-threshold depolarizing stimuli they are accompanied by active changes of potential: by a local response and an action potential, connected with changes in the ionic permeability of the membrane.

Passive Changes of Membrane Potential

In its passive electrical characteristics, the nerve fiber usually resembles a conductor with a core or a telephone cable placed in a conducting medium (Hermann, 1879, 1905; Cremer, 1900, 1932; Cole, 1928, 1932, 1964; Hodgkin and Rushton, 1946; Taylor, 1963, etc.; see also Chailakhyan, 1962).

37

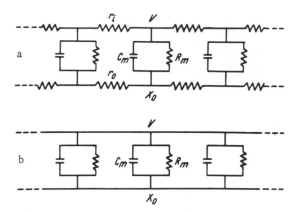

Fig. 11. Equivalent circuit of passive electrical properties of an axon under different experimental conditions. a) Circuit for the case in which the regions of
a membrane are connected together by the resistances
of the internal axoplasm (r_i) and of the external solution (r_0); C_m, capacitance; R_m, resistance of the membrane; b) axon shunted along its length by an axial wire.
External resistance is low because the axon is immersed
in a large volume of fluid.

The equivalent circuit (electrical model) of such a cable is
shown in Fig. 11a. Here each element of the excitable membrane
is represented by a cell consisting of a capacitor (C_m) and re-
sistor (R_m) connected in parallel. The cells are connected to each
other by the resistances of the axoplasm (r_i) and the external solution (r_0). It is taken that r_i and $r_0 \ll r_m$.

At the point x_0 an electrode connected to a dc source is applied to the fiber. A second electrode, not shown in the diagram,
is placed on the cable at a considerable distance from the first.

If the current passed only through one RC segment, the potential V on the plates of the capacitor C would increase exponentially,
in accordance with the law

$$V = IR_m \left(1 - e^{-t/R_m C_m}\right). \tag{8}$$

The presence of neighboring RC segment in the cable (or
fiber), connected in parallel through the resistances r_i and r_0,
substantially changes the development of the electronic potentials.

At the point of application of the electrode to the fiber (x_0) the amplitude of the electrotonic potential is reduced because of branching of the current along the fiber. The final level of V_0 (the "saturation potential"), to which the potential rises at the point x_0 for the given current density is determined, not simply by the resistance of the membrane itself R_m, but by the input resistance R_{in} of the fiber at this point, i.e., by the overall resistance between the external solution and the axoplasm.

The input resistance is dependent on the resistance of the membrane R_m, the specific resistance of the axoplasm R_i, and the diameter of the fiber d:

$$R_{in} = \frac{1}{\pi} \sqrt{\frac{R_m R_i}{d^3}} \tag{9}$$

Besides reducing the amplitude of the electrotonic potential at the point x_0, branching of the current also changes the development of this potential in time.

In a cable, unlike what takes place during the passage of a current into one segment, the potential at the point of application of the electrode does not rise exponentially [Eq. (8)], but more steeply, in accordance with the law

$$V_x = V_\infty \, \text{erf} \sqrt{\frac{T}{\tau'}}, \tag{10}$$

where erf is the probability integral and τ' is the time during which the potential increases to 84% of its final value [erf(1) = 0.8425].

However, the most important results of branching of the current in a cable are changes in the potential in segments of the fiber next to the point of application of the electrode. Experiments and theoretical calculations show that the current density and potential shift are greatest actually at the point of application of the current to the fiber. With an increase in the distance from this point, the current density and potential shift diminish exponentially.

In the steady state,

$$V_x = V_0 e^{-x/\lambda}, \tag{11}$$

where V_x is the potential at the point x, V_0 the potential actually under the electrode (at the point x = 0); and λ the constant of length, i.e., the distance in which the electrotonic potential falls by e times. This distribution of potential along the fiber is called electrotonic.

The changes in membrane potential of a lobster nonmedullated nerve fiber at different distances from the point of application of a steady cathodal current to it, observed experimentally and calculated by Eq. (11), are shown in Fig. 12a.

Clearly the higher the resistance of the membrane compared with the resistances of the axoplasm and external solution, the greater the length constant λ. Calculations showed that

$$\lambda = \sqrt{\frac{r_m}{r_i + r_0}} . \tag{12}$$

where r_m, r_i, and r_0 are the resistances of the membrane, axoplasm, and external solution (respectively) per unit length of fiber.

If r_0 is small by comparison with r_i, it can be ignored. In that case

$$\lambda = \sqrt{\frac{r_m}{r_i}} . \tag{12a}$$

The value of λ can also be expressed in terms of r_m and r_i. In that case

$$\lambda = \frac{1}{2} \sqrt{\frac{dR_m}{R_i}} . \tag{12b}$$

The principles governing distribution of the current and potential in a nerve fiber are also applicable to a nerve cell with branching processes.

Rall (1964) showed that the dendritic tree of a neuron is equivalent in its passive electrical properties to a passive unbranched membrane cylinder of infinite length.

An electric current passing through the body of a neuron flows partly through the soma membrane and partly through the processes. The passive changes in membrane potential thus arising are determined by the specific contributions made by the cell body and its processes. To define these contributions, Rall (1964) suggested

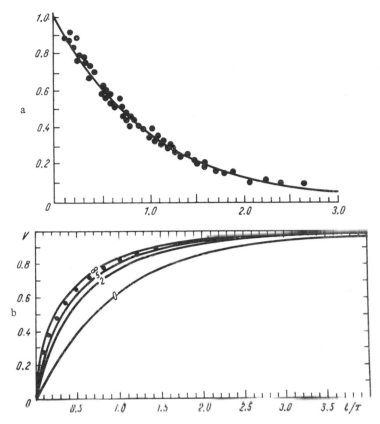

Fig. 12. Electrotonic changes in membrane potential: a) steady-state distrbituion of potential in extrapolar region of isolated crab nerve fiber. Ordinate: potential in multiples of value measured directly under the cathode. Abscissa: distance from cathode in relative units (Hodgkin and Rushton, 1946); b) passive change in soma membrane potential of theoretical neuron in response to application of a steady-current. Time (t) expressed in units equal to time constant (τ) (abscissa). Electrotonic potential (ordinate) expressed relative to final steady value. Values of the dendritic dominance factor (see text) are shown on the curves (Rall, 1964).

the parameter of "dendritic dominant" ρ, the ratio between the total dendritic conductance of the membrane and the conductance of the soma membrane.

If the dendrites are completely dominant ($\rho = \infty$), the potential on the cell soma membrane rises in accordance with the same law as the potential in the nerve fiber (curve "∞" in Fig. 12b).

If, on the other hand, there are no dendrites ($\rho = 0$), the potential rises exponentially [Eq. (8); curve "0" in Fig. 12b], since in a spherical cell all regions of the membrane are connected in parallel and, consequently, the electrical circuit of the cell coincides with the circuit of its membrane (Taylor, 1963).

The intermediate curves in Fig. 12b correspond to different values of the dendritic dominance factor ρ.

Changes in the membrane potential with time are similar in character for most neurons. It is only in giant neurons of mollusks (Aplysia, Helix pomatia, etc.) and neurons of the spinal ganglia, which have no dendrites, that the changes in membrane potential caused by pulses of steady current are virtually exponential in character (Tauc, 1962a,b; Gerasimov, Kostyuk, and Maiskii, 1965b).

The cable properties of nerve and muscle fibers have a significant effect not only on the development of the electrotonic potential, but also on the character of the active responses: the threshold, amplitude, steepness of rise, and duration of the action potential (Chapter VIII). Accordingly, in order to differentiate between the changes in these parameters associated with the cable properties of the fiber, i.e., with the geometry of the excitable structure, and changes determined by the properties of the excitable membrane itself, in experimental work it is customary to use so-called "space-clamp" methods.

The best method for use with mollusk giant axons is to introduce a long, low-ohmic, metal electrode into their axoplasm (Marmont, 1949). Such an axial electrode shunts the resistance of the internal contents of the fiber and thus makes all points of the short-circuited region of the membrane equipotential (Fig. 11b).

If the current is passed through such an intracellular electrode and another concentric electrode placed in the solutions surrounding the fiber, spreading of the current along the fiber will be eliminated, for the resistance of the metal electrode is many times less than the resistance of the axoplasm of neighboring regions of the fiber and all points of the membrane in the axon segment investigated will undergo identical changes at the identical time.

The action potential arising in such a uniform area of the membrane is usually called the "membrane action potential," to distinguish it from the "propagated action potential" developing in cases when the current spreads along the cable.

Another version of the "space clamp" method of obtaining a uniform membrane is to place a nonmedullated nerve fiber on two sucrose bridges (Stämpfli, 1954) isolating the segment of the nerve fiber membrane to be investigated from the rest of the membrane (for details of this method, see Julian, Moore, and Goldman, 1962; Guttman and Barnhill, 1966).

In the medullated nerve fibers of vertebrates the myelin sheath effectively isolates the excitable membrane of the tested node of Ranvier from the membrane of the next node. The area of the membrane of each node is so small (about 30 μ^2) that there are good grounds for regarding the potential change in this region as sufficiently uniform, even if this membrane is maximally active (see Dodge, 1963). To study generation of the membrane action potential in medullated nerve fibers, the single node of Ranvier technique suggested by Tasaki and Takeuchi (1942) and subsequently improved by Frankenhaeuser (1957a) is therefore used. The improvements made by Frankenhaeuser to this method consist essentially of the creation of an electronic negative feedback system, automatically eliminating the longitudinal current in the internodal segment of the fiber. In this way, the membrane potential of the tested node can be measured directly by means of extracellular electrodes.

The Local Response

Passive, electrotonic changes in membrane potential induced by a depolarizing current generate an active, subthreshold electrical response known as the local response if this current approaches the threshold strength.

Historical

The first indirect evidence of the existence of the local response was obtained by Katz in 1937 when he compared changes in the threshold of excitability of a frog nerve at short time intervals after the beginning of action of cathodic and anodic stimuli of different strengths. The following year, Hodgkin (1938) obtained direct evidence of the generation of a local response in isolated nerve fibers of the crab <u>Carcinus</u> <u>maenas</u>. This work by Hodgkin laid the foundations of modern knowledge of the local response and its relationships with the electrotonic potential and action potential.

Within a very short time local responses had been recorded in other nonmedullated nerve fibers also; the squid giant axon (Pumphrey, Schmidt, and Young, 1940), the nerve fibers of the walking leg of the lobster (Hodgkin and Rushton, 1946).

Although the first facts supporting the existence of local responses were obtained by Katz working with frog nerves, attempts to record local responses directly in medullated nerve fibers for a long time proved unsuccessful (Blair and Erlanger, 1936; Blair, 1938; Lorente de Nó, 1947). As later investigations showed, these failures were due to the fact that the experiments to record these potentials were carried out on nerves covered with a sheath which, together with its connective tissue layers and its intercellular fluid, shunted the weak electrical responses of the nerve fibers.

This technical difficulty was overcome by Katz (1947), who by means of an ingenious method balanced the passive polarization changes arising in the nerve under the cathode of the applied current, and thus was able to differentiate between local responses and electrotonic potentials.

A short time later, Rosenblueth and Luco (1950) and Rosenblueth and Ramos (1951) published convincing evidence of existence of a local response in medullated nerve fibers of the spinal roots of the cat. In the same year Schoepfle and Erlanger (1951) recorded a local response in the nerve fibers of a frog phalangeal preparation.

Although these investigations were not yet at a sufficiently high technical level, because global recordings were obtained from groups of fibers, they nevertheless yielded important evidence to show that the local response arises in intact nerve fibers also and is not an abnormal phenomenon due to injury in the course of microdissection and other special manipulations.

However, in order to study the properties of the local response in the nerves of vertebrates in more detail, the experiments had to be carried out on isolated nerve fibers, and the investigation had to be undertaken in such a way that all side-effects distorting the dynamics of development of the potentials at the point of stimulation of the fiber were eliminated as far as possible.

These conditions were best satisfied by the single node of Ranvier preparation. By the use of this preparation, Huxley and Stämpfli (1951) and Castillo and Stark (1952) accurately recorded local responses in a medullated nerve fiber and showed that these responses are indistinguishable, in principle, in their properties from the local responses studied by Hodgkin in nonmedullated fibers.

A local response was subsequently recorded in skeletal muscle fibers (Katz, 1947; Jenerick, 1956; Knuttson, 1964) and nerve cells (Tauc, 1955; Coombs, Curtis, and Eccles, 1959; Kostyuk and Semenyutin, 1961; Sasaki and Oka, 1963; Kawai and Sasaki, 1963; Arvanitaki and Chalasonitis, 1965; Gerasimov, Kostyuk, and Maiskii, 1965a,b) during stimulation of the fibers via intracellular microelectrodes. However, it should be noted that because of the complex character of the neuronal action potential (consisting of "initial" and "soma-dendritic" components), it is often difficult to detect a local response in the cell body. Moreover, if the neuron is functioning properly, the interval between strength of stimulation at which the first signs of the local response appear and at which the local response changes into an action potential is very narrow. If the cell is injured, this gap widens and the local response can be clearly detected during stimulation at subthreshold strength.

It is a very interesting fact that the local response in a nerve cell arises not only to artificial stimulation, but also to natural synaptic stimulation. In this case the local response is superposed on the excitatory postsynaptic potential and causes it to rise more steeply to the threshold level (Sasaki and Oka, 1963; Arvanitaki and Chalasonitis, 1965).

Method of Detecting the Local Response

In order to differentiate between the local response and the passive electrotonic potential, special methods are usually employed.

Pulses of depolarizing and hyperpolarizing current of identical strength are applied to the same point of a nerve or muscle fiber.

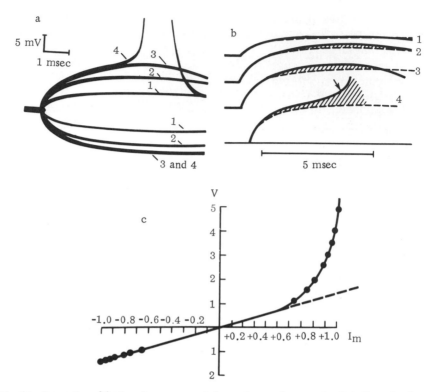

Fig. 13. Properties of the local response: a) change in membrane potential of a node of Ranvier under the influence of pulses of depolarizing and hyperpolarizing currents of subthreshold (1, 2, 3) and threshold (4) strength; b) graphic analysis of ratios between passive and active components of membrane depolarization. Potential changes beneath the anode are shown by a broken line, and beneath the cathode by a continuous line. 10°C (Khodorov, 1965a); c) change in membrane potential as a function of stimulus strength. Duration of pulses of current 0.05 msec. Abscissa: strength of current (depolarizing indicated by plus sign, hyperpolarizing by minus sign) in multiples of threshold, taken as unity; ordinate: changes of potential measured 0.3 msec after application of stimulus, in arbitrary units (Castillo and Stark, 1952).

The strength of these pulses is gradually increased from test to test up to the threshold level. The catelectrotonic and anelectrotonic changes in membrane potential recorded in response to currents of the same strength are compared by graphic subtraction. It is assumed that anodic (hyperpolarization) changes of membrane potential are purely passive in character.

An example of this method of analysis is illustrated in Fig. 13a,b. In this case a single node of Ranvier is stimulated by pulses of steady current of long duration. If the current is weak, the changes of potential evoked by anodic and cathodic pulses are about equal in magnitude, for the resistance of the membrane remains unchanged during passage of the current. As the current increases in strength, the depolarization changes of potential start to rise more and more above the corresponding hyperpolarization changes. This disturbance of symmetry is evidence of the development of an active local response (the shaded sector in Fig. 13b).

Local responses are found not only during the action of pulses of current of long duration, but also when short stimuli are applied. In the latter case, however, the local responses appear in the phase of decline of the electrotonic potential.

It will be seen in Fig. 14a that after a short anodic pulse the potential smoothly returns to its initial level. By contrast, after a subthreshold cathodic stimulus the potential begins to fall after a delay due to generation of an additional depolarization wave — the local response.

It must be emphasized, however, that the anelectrotonic changes in membrane potential are virtually passive in character only within a certain range of strength and durations of the hyperpolarizing current. Strong and prolonged hyperpolarization of the membrane causes an appreciable change in resistance and, consequently, in the time constant of the membrane, as a result of which the dynamics of development of the anelectrotonic potential begins to deviate from that expected theoretically (Ooyama and Wright, 1961; Grundfest, 1964; see also page 282).

Properties of the Local Response

The properties of the local response have been studied from two aspects. First, its differences from the electrotonic potential have been demonstrated, and second, the local response has been compared with the action potential. The local response differs very considerably in its properties from the passive electrotonic potential: 1) whereas the amplitude of the electrotonic potential is directly proportional to the current strength, the local response is a nonlinear function of this strength (Fig. 13c) (Hodgkin, 1938; Castillo and Stark, 1952); 2) during stimulation of a uniform membrane, the electrotonic potential, as was mentioned above, rises and falls on an exponential curve at a rate determined, for a current of this strength, by the RC of the membrane. By contrast, the curve re-

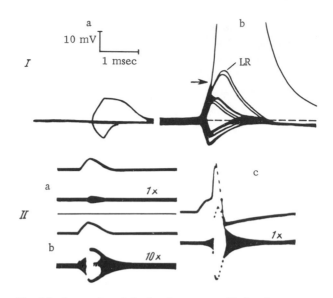

Fig. 14. Properties of the local response. I) Local responses
in node of Ranvier of a frog nerve fiber, evoked by short
pulses of depolarizing current. Changes of potential in response to the action of hyperpolarizing pulses of the same
strength are given for comparison; a and b represent different preparations. Calibration the same throughout. LR,
local response. Arrow indicates stimulus off (Khodorov and
Belyaev, 1963a). II) Changes in impedance of squid giant
axon membrane during subthreshold (a and b) and threshold
(c) activity. Records a and b were made with subthreshold
stimulation of approximately equal strength, but in record
c the changes in impedance are magnified by ten times
for the sake of increased clarity (10x). Top curves show
changes in membrane potential, local response (a and b),
and action potential (c) (Tasaki, 1950a).

flecting the rise of the local response is S-shaped. In the case of
the action of very short stimuli, it can be seen that the local response continues to rise for some time after the end of the stimulus evoking it (Fig. 14a,b) (Hodgkin, 1938; Khodorov and Belyaev, 1963); 3) the development of the electrotonic potential with
time is independent of the strength and duration of the current,
yet the local response reaches its maximum last of all, as the
stimulus approaches the threshold strength (Hodgkin, 1938; Hodgkin and Huxley, 1952d).

The only similarity between the local response and electro-tonic potential is that both show a graded dependence on stimulus strength (admittedly, as was pointed out above, this dependence differs: it is linear in one case, but nonlinear in the other) and that during the local response, just as during the development of the catelectrotonic potential, the excitability of the nerve fiber rises (Katz, 1937; Pumphrey, Schmidt, and Young, 1940; Hagiwara and Oomura, 1958).

In some of its properties the local response resembles the action potential: 1) like the action potential it can develop without maintained current — rising at the beginning, then falling — after the end of the short stimulus evoking it (Fig. 14b) (Hodgkin, 1938); 2) during the local response, just as during spike generation, the impedance of the membrane falls, i.e., its ionic permeability rises; the differences between the changes in impedance in the two cases are purely quantitative (Fig. 14b,c) (Tasaki, 1959); 3) in the after-period of the local response there is a phase of decreased excitability, which is very similar in its dynamics with the period of relative refractoriness accompanying the action potential (the "local relative refractory phase" of Pumphrey, Schmidt, and Young, 1940).

On the other hand, unlike the action potential, the local response: 1) has no precise threshold: it is obtained in a zone of subthreshold strengths of stimulation, 0.5-0.9 of the threshold (Hodgkin, 1938; Castillo and Stark, 1952); 2) it is not accompanied by absolute refractoriness but, on the contrary, during the local response the level of excitability usually rises (see above); 3) the local response can undergo summation if a second subthreshold stimulus is applied against the background of a local response to a preceding stimulus (Hagiwara and Oomura, 1958; Pumphrey, Schmidt, and Young, 1940); 4) it does not obey the "all or nothing" rule: its amplitude and steepness of rise increase gradually both as the strength of the subthreshold current is increased (Hodgkin, 1938) and in response to a regularly repeated stimulation of the nerve fiber (node of Ranvier) by subthreshold stimuli of low frequency (Kitamura, 1961).

However, these clear boundaries between the action potential and local response are partly or completely obliterated during the action of various factors inhibiting regenerative depolarization of

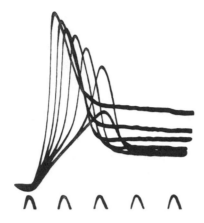

Fig. 15. Graded action potentials of a node of Ranvier after preliminary application of a subthreshold depolarizing current. Stimulation by steady current of different strength. Time marker 1 msec (Müller, 1958).

the membrane (Müller, 1958; Uchizano, 1960; Uehara, 1960; Ichioka, Uehara, and Kitamura, 1960; Khodorov, 1962; Belyaev, 1963, 1964; Kirzon and Chepurnov, 1963). For instance, during weak catelectrotonus or during incomplete narcosis by the action of excess K^+ ions, procaine, urethane, anisotonic solutions, eserine, neostigmine, or other agents, the fiber loses its ability to generate normal action potentials. Instead, it generates responses which occupy an intermediate position in their properties between the action potential and the local response (Fig. 15).

Role of the Local Response in Critical Membrane

Depolarization

The local response plays an important role in the critical depolarization of the membrane produced by stimulation.

As Fig. 16a shows, during threshold stimulation the passive, electrotonic potential and the local response (the shaded sector) are added together, and the action potential arises when the sum of these two components reaches the critical level.

According to the observations of Hagiwara and Oomura (1958), in squid giant axons the local response to a current of rheobase strength accounts for between 20 and 30% of the total threshold change in membrane potential. In nodes of Ranvier, this contribution varies from 20 to 40% and, under some conditions, even to 50% (Khodorov and Belyaev, 1963a; Khodorov, 1965a).

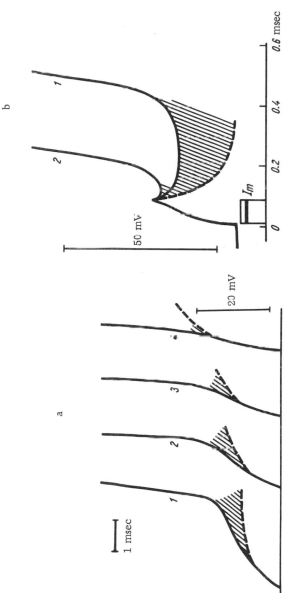

Fig. 16. Contribution of local response to threshold change of membrane potential arising during stimulation of different strengths and duration. a) Steady current of 1, 1.5, 2, and 3 times rheobase strength (curves 1-4, respectively) Khodorov and Belyaev, 1963a); b) short stimuli of 1 and 1.5 times rheobase strength (curves 1 and 2, respectively). The shaded sector shows the local response measured by graphic subtraction of the anodic change of potential from the cathodic. I_m denotes stimulus applied.

This fact is of interest from the theoretical point of view. When the adaptive properties of excitable tissue are described, it is usual to emphasize only its ability to "escape from the stimulus," as is manifested by accommodational increase in the threshold of stimulation in the event of a decrease in the gradient or an increase in the duration of action of the current (page 191).

However, as we shall see, during the action of a stimulus, besides accommodation, a process with the opposite sign, facilitating critical depolarization of the membrane, also develops in the excitable tissue. This process is a local subthreshold excitation, which facilitates spike generation in response to a weaker (by $\frac{1}{5}-\frac{1}{2}$) current than would otherwise be necessary to cause excitation if the critical depolarization were entirely and completely attained through a passive change in membrane potential.

The local response plays a particularly important role in critical membrane depolarization during stimulation of excitable structures by very short stimuli. In this case the action potential arises only after the end of the threshold pulse of current in the waning phase of the electrotonic potential, and the contribution of the latter to the critical change of potential is very small (Fig. 16b).

With an increase in stimulus strength the ratio between the passive and active components of the prespike depolarization alters in favor of the first, during the action of both long (Fig. 16a) and short (Fig. 16b) pulses of current.

The reasons for this phenomenon will become clear after we have examined the changes in ionic permeability and ionic currents during critical depolarization of the excitable membrane (Chapter VI).

Chapter IV

The Action Potential

Historical

Bernstein (1902, 1912), the founder of the modern membrane theory, regarded the action potential as the "negative wave" of the resting potential. He considered that during stimulation of a nerve or muscle fiber certain chemical changes which give rise to a reversible increase in the ionic permeability of the cell membrane take place at the point of application of the stimulus. The membrane loses its selective permeability of K^+ ions, as the result of which the potential difference which exists at rest is quickly equalized, and the resting potential falls to zero. Subsequent restoration of the original ionic permeability leads to a return of the membrane potential to the level of the resting potential.

After Bernstein's investigations, many attempts were made to study changes in ionic permeability of membranes during excitation experimentally. Lullies (1930), followed by Cole and Curtis (1936), for instance, investigated the longitudinal impedance of the frog's sciatic nerve and showed that it decreases during the conduction of nervous impulses. Similar changes were obtained by Curtis and Cole (1938) for the transverse impedance of the squid giant axon. However, in these investigations the measuring current was used simultaneously for the purpose of stimulation, so that the observed changes in impedance could not be analyzed with sufficient thoroughness. Accordingly, in 1939 Cole and Curtis made a special investigation of this problem on squid giant axons. They measured impedance by means of an alternating current, the frequency of which was varied within wide limits. During

53

passage of the pulse between the impedance electrodes (the nerve was stimulated via electrodes placed on one of its ends) the balance of the bridge was upset as a result of the fall in impedance, and this was shown by the fan-shaped broadening of the recorded line (Figs. 14b,c and 79a). These measurements and their analysis by the clock diagram method (Cole, 1932) showed that during action potential generation the capacitance of the membrane alters by not more than 2%, whereas its ionic conductance is increased by about 40 times (its ohmic resistance falls from 1000 to 25 $\Omega \cdot cm^2$).

These investigations, which were subsequently confirmed by Tasaki (1953) on the single node of Ranvier of frog nerve fibers, gave direct proof of Bernstein's hypothesis that ionic permeability of the membrane is increased on excitation.

However, one question remained unanswered: to what ions in particular is this permeability increased? As was mentioned above, Bernstein considered that during excitation there is an increase in the permeability of the membrane to all ions, and he therefore regarded the action potential purely as a transient disappearance of the resting potential. The discovery that the action potential is greater in magnitude than the resting potential upset this simple idea.

The first indirect indications of the existence of a difference between the magnitudes of the resting potential and action potential were obtained in experiments with extracellular electrodes. Schäfer (1936), for instance, measured the membrane potential from the potential difference between intact and injured regions of the gastrocnemius and sartorius muscles of a frog and found that the action potential is 15-20 mV greater than the resting potential.

Similar results were obtained with the use of extracellular electrodes at the beginning of 1939 by Hodgkin and Huxley in experiments on the squid giant axon. The resting potential measured in these experiments was 37 mV, while the action potential was −73 mV (cited by Hodgkin, 1964).

However, it was impossible, by means of extracellular electrodes, to measure the absolute values of the membrane potential because of the shunting effect of the thin layer of fluid covering the nerve or muscle between the injured and intact segments, and the last word had thus to be postponed until direct measurements of the membrane potential could be made by means of intracellular electrodes.

Such measurements were first made almost simultaneously on squid giant axons by Hodgkin and Huxley (1939) and by Cole and Curtis (1939).

These workers introduced a thin metal electrode inside an axon and used it to show that during generation of the nervous impulse the resting potential does not just simply disappear, as Bernstein supposed, but a new potential difference of opposite sign is created. Under these circumstances, the amount by which the action potential exceeds the resting potential in magnitude, the so-called overshoot, is more than 40 mV.

Development of the method of measuring the membrane potential by means of glass intracellular microelectrodes (Graham and Gerard, 1946; Ling and Gerard, 1949) confirmed this conclusion for other excitable structures (Nastuk and Hodgkin, 1950). It was shown that in virtually all properly functioning excitable cells the action potential is 40-50 mV greater than the resting potential.

Medullated nerve fibers are quickly damaged by the insertion of a microelectrode, and for this reason, in order to measure absolute values of the resting and action potentials in these excitable structures, a special technique was developed so that these potentials could be recorded from a single node of Ranvier with the aid of ordinary extracellular electrodes (Huxley and Stämpfli, 1951a; Frankenhaeuser, 1957a) (see page 43).

Investigations showed that the resting potential of the node of Ranvier of a carefully dissected nerve fiber is about − 70 mV (Dodge, 1963, gives the value −75 mV), while the amplitude of the action potential may reach 115-120 mV.

Some physiologists have interpreted all these facts as evidence of a crisis in the membrane theory of biopotentials (see Nasonov, 1949, 1959). In reality, as time has shown, these discoveries did not lead to the abandonment of the membrane theory, but, on the contrary, they stimulated its further development.

Effect of Changes in External and Internal Sodium and Potassium Ion Concentrations on Action Potential Generation

It was shown by Overton as long ago as in 1902 that skeletal muscles lose their excitability in an isotonic solution containing less than 10% of the normal concentration of NaCl. Under these conditions, as experiments showed, chloride ions are not an essential component of the Ringer's solution: excitability still remained if the sodium chloride was replaced by sodium nitrate, bromide, sulfate, phosphate, bicarbonate, benzoate, and so on. But of all the cations which have been tested, only lithium was found to be suitable as a substitute for sodium.

Overton attempted to reproduce all these effects on the frog nerve. However, this tissue remained excitable for a long time in a sodium-free solution, and as we now know, this was because of difficulty in washing out the whole of the sodium from the intercellular spaces.

Direct proof that Na^+ ions are essential for generation of the nervous impulse was first obtained in 1936 by Kato in experiments on isolated medullated nerve fibers, and later by Erlanger and Blair (1938) on the sensory spinal roots of the frog and by Katz (1947) on crustacean nerve fibers.

The view of Overton (1902) that Na^+, but not Cl^-, ions are important to conduction was confirmed by Lorente de Nó (1940, 1947), who replaced the NaCl in Ringer's solution by choline chloride.

Not all these results could be explained by Bernstein's theory, and they led Hodgkin and Huxley to suggest that the nerve fiber membrane, in a state of excitation, does not simply lose its selective ionic permeability, as Bernstein thought, but instead of being selectively permeable to K^+ ions it becomes selectively permeable to Na^+ ions.

Preliminary calculations showed that an overshoot of about 60 mV can be obtained if it is accepted that the intra-axonal Na^+ ion concentration is approximately one-tenth of the Na^+ ion concentration in the medium. A special investigation on squid axons was conducted by Hodgkin and Katz (1949) to test this hypothesis. They varied the external Na^+ ion concentration within wide limits and investigated the resulting changes in amplitude and shape of the spreading action potential. Their experiments showed that within a certain range of sodium concentrations (as the Na^+ concentration fell to 50-70% of its initial value or was increased by 50%), the potential difference on the active membrane behaves as an approximately linear function of the logarithm of $[Na]_0$ (Fig. 17a). This accorded well with the hypothesis that the active membrane of the nerve fiber behaves as a sodium electrode, for which the equilibrium potential is given by

$$E_{Na} = \frac{RT}{F} \ln \frac{[Na]_i}{[Na]_0} = 58 \log \frac{[Na]_i}{[Na]_0}, \quad \text{at } 18° C \tag{13}$$

The maximal rate of rise of the action potential within this concentration range was approximately proportional to the sodium concentration in the external solution.

Only when the Na^+ ion concentration in the external solution fell to 20-30% of its initial level in seawater was there a disproportionately large decrease in the amplitude and in the steepness of rise of the action potential.

The action potential of the squid giant axon is known to be accompanied by a well-marked after hyperpolarization of the membrane, reaching 5-15 mV. The experiments of Hodgkin and Katz (1949) showed that the amplitude of this after-potential is independent of the Na^+ ion concentration in the solution, but rises or falls distinctly in response to even a slight decrease or increase in the external K^+ ion concentration. In one of their experiments, for instance, removal of K^+ from the seawater solution increased the after-hyperpolarization to 23 mV, while doubling the K^+ concentration (from 10 to 20 mM) lowered this potential to 7 mV. Hodgkin and Katz explained all these facts by reference to the "constant field equation" (7).

At rest, as was pointed out above, we find the ratio

$$P_K : P_{Na} : P_{Cl} = 1.0 : 0.04 : 0.45.$$

To obtain an action potential with an amplitude of 100 mV, how-

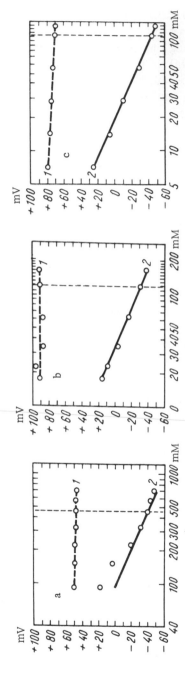

Fig. 17. Relationships between sodium concentration in external solution and potential difference on resting and active membrane (after Hodgkin, 1951): a) squid giant axon; b) fiber of frog sartorius muscle; c) single node of Ranvier from frog nerve fiber. Abscissa, Na^+ ion concentration in solution, logarithmic scale [vertical broken line indicates Na^+ ion concentration in sea water (a) or in Ringer's solution of normal composition (b, c)]. Ordinate, transmembrane potential difference at rest (1) and at summit of action potential (2). Continuous lines drawn at a gradient of 58 mV for a ten-fold change in Na^+ ion concentration.

ever, the sodium permeability of the membrane must be increased to 500 times: from 0.04 to 20.

With this ratio between the coefficients of permeability, Hodgkin and Katz were able to describe the changes in the action potential of an axon observed during variation of the external Na^+ ion concentration between 465 and 324 mM. However, with a greater decrease in $[Na]_0$ — to 227 mM, and, in particular to 91 mM — the decrease in amplitude of the action potential observed experimentally was much greater than that expected theoretically. The impression was obtained that under these circumstances the maximum to which P_{Na} rose during depolarization of the membrane was reduced.

These experiments thus confirmed the view that generation of the action potential is connected with an increase in the sodium permeability of the membrane. These results also compelled investigators to seek the causes of restoration of the initial potential difference during excitation and the mechanism of genesis of after-hyperpolarization.

To explain these phenomena it was suggested that the increase in sodium permeability of the membrane is only a transient process, to be followed by inactivation of the mechanism responsible for the transport of sodium ions through the membrane down the electrochemical gradient, and that depolarization of the membrane also causes a gradual increase in potassium permeability.

Inactivation of sodium permeability is the main cause of the absolute refractory state of the nerve fiber during the action potential. The increase in potassium permeability, with the consequent increase in outward potassium current, accelerates the restoration of the membrane potential, and also contributes to refractoriness.

The validity of this conclusion was subsequently confirmed by the elegant experiments of Hodgkin and Keynes (1955) on giant axons of the cuttlefish (Sepia officinalis) in which the intracellular concentration of K^+ ions was determined immediately after the experiment by activation analysis. To eliminate the effect of active ion transport on the results, this process was blocked by 0.2 mM dinitrophenol.

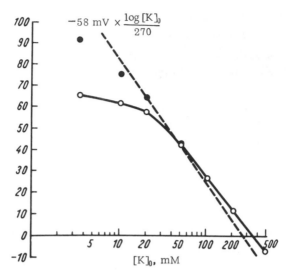

Fig. 18. Effect of external potassium concentration on resting potential and after-hyperpolarization of the cuttle-fish giant axon membrane. Active transport blocked by 0.2 mM dinitrophenol. Abscissa, K^+ concentration in mM (logarithmic scale); ordinate, mean values of trans-membrane potential difference at rest (unfilled circles) and during phase of after-hyperpolarization (filled cir-cles). K^+ concentration measured in this axon after ex-periment, 270 mM. Temperature 17-19°C (Hodgkin and Keynes, 1955b).

The curve showing the changes in resting potential and membrane potential in the after-hyperpolarization phase as functions of the logarithm of the K^+ ion concentration in the solution is given in Fig. 18. Curve 1 is indistinguishable from the corresponding curve obtained for an axon untreated with dinitrophenol. The slope of the curve at high values of $[K]_0$ was ~50 mV for a tenfold change in $[K]_0$, but it decreased at low values of $[K]_0$. As was pointed out above, this indicates that during strong depolarization of the membrane resulting from the high $[K]_0$, the greater part of the current is carried by K^+ ions (Hodgkin and Huxley, 1953). Other ions, however, make an important contribution to the resting potential only at low values of $[K]_0$.

The slope of the curve of membrane potential versus $\log [K]_0$ was steeper in the phase of after-hyperpolarization (filled circles). This indicates that during the action potential the potassium permeability rises, and it does not return for several milliseconds to its initial level. At $[K]_0 = 52$ mM, the membrane potential in the phase of after-hyperpolarization was equal to the potassium equilibrium potential obtained for that particular axon by direct determination of the K^+ concentration in the cytoplasm by activation analysis (see page 65). The broken line was obtained from the Nernst equation for the equilibrium potential at different values of $[K]_0$.

Important results confirming the hypothesis of the role of Na^+ and K^+ ions in action potential generation were subsequently obtained in experiments with internal perfusion of squid giant axons.

Baker, Hodgkin, and Shaw (1962), in their experiments, replaced isotonic K_2SO_4 solution by a mixture of isotonic solutions of K_2SO_4 and Na_2SO_4. These experiments showed that an increase in the internal Na^+ concentration leads to a reversible decrease in the overshoot of the action potential. The relationship between the values of the membrane potential at the peak of the spike (E_s), the resting potential (E_r), and the potential at the maximum of the after-hyperpolarization (E_a) and the Na^+ and K^+ ion concentrations in the solution perfusing the fiber is shown in Fig. 19. The points correspond to experimental data, while the curves were calculated from the reduced constant field equation

$$E = \frac{RT}{F} \ln \frac{[K]_0 + b\,[Na]_0}{[K]_i + b\,[Na]_i} \tag{7a}$$

where $[K]_0$ and $[Na]_0$ are the activities of the ions; R, T, and F have their usual meaning; $b = P_{Na}/P_K$; E is the potential on the inside of the membrane.

In this case (Fig. 19), satisfactory agreement between the experimental and theoretical results was obtained for the following values of the coefficient b: for the active membrane (at the peak of the spike) 7, in the resting state 0.08, and at the height of after-hyperpolarization 0.03. When the amplitude of the action potential reached 120 mV, the ratio P_{Na}/P_K at the peak of the spike rose to 14, compared with its values of 0.06 at rest and 0.03 during after-hyperpolarization.

It follows from the equation that if $[Na]_i$ is small compared with $[K]_i$, the overshoot of the action potential is limited by the in-

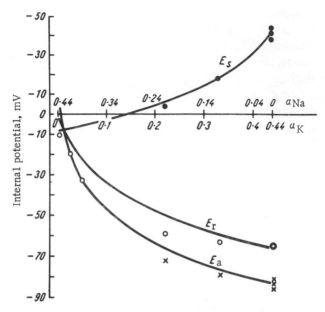

Fig. 19. Effect of replacement of K^+ by Na^+ in internal solution on active membrane potential (at the peak of the spike) (E_s), resting potential (E_r), and after-hyperpolarization (E_a). Abscissa, activity (a) of potassium and sodium (in moles/liter) in internal solution (scale divisions are read from left to right for K, from right to left for Na); ordinate, inside membrane potential in mV. The smooth curves were drawn according to Eq. (7a) for $P_{Na}/P_K = 7$ for top curve, for $P_{Na}/P_K = 0.08$ for middle curve, for $P_{Na}/P_K = 0.03$ for bottom curve. External solution was sea-water in which: $a_{Na} = 0.32$ and $a_K = 0.0068$ mole/liter.

ternal K^+ ion concentration only. When $[Na]_i = 0$ and $[Na]_0 \gg [K]_0$ the equation reduces to

$$E = \frac{RT}{F} \ln \frac{b [Na]_0}{[K]_i}. \tag{7b}$$

It thus follows that in the absence of Na^+ ions in the internal solution, a decrease in $[K]_i$ must lead to an increase in the overshoot. The experiments confirmed the theoretical predictions. Replacing half the K_2SO_4 in the internal solution by glucose, leading to a decrease in K^+ ion activity by 40%, increases the overshoot by about 10 mV. After sixfold dilution of the K_2SO_4 with glucose the overshoot was increased by 30 mV.

The experiments with internal perfusion of the axon also confirmed the hypothesis that a significant increase in potassium permeability of the membrane takes place in the descending phase of the action potential and that an increase in the outgoing current of these ions accelerates repolarization of the membrane. On replacement of K^+ in the internal solution by cesium or rubidium ions or after dilution of the internal K_2SO_4 or KCl solution with glucose or sucrose, the repolarization phase is prolonged and after-hyperpolarization of the membrane disappears. I shall return to the examination of these effects in more detail later.

The results of experiments in which the internal concentrations of Na^+ and K^+ ions in the solution were changed thus led to the conclusion that differences in the concentration of these ions on the two sides of the membrane are the immediate source of energy for the nervous impulse and that the action potential arises as a result of changes in ionic permeability which had previously been deduced from the results of experiments with intact axons.

The view of Hodgkin, Huxley, and Katz on the role of sodium ions in action potential generation has been tested in experiments on various excitable structures.

Nastuk and Hodgkin (1950), working with skeletal muscle fibers (Fig. 17b), and Huxley and Stämpfli (1951a), with single nodes of Ranvier from frog nerve fibers (Fig. 17c) showed that in these excitable structures, just as in squid giant axons, the magnitude of the overshoot is a linear function of the logarithm of Na^+ ion concentration in the solution. The gradient of the E versus log $[Na]_0$ line was close to the theoretically expected value of 58. These experiments also confirmed that the maximum rate of rise of the action potential is approximately proportional to the external sodium ion concentration.

Similar results were obtained by Dalton (1958) in experiments on giant axons of the lobster (Homarus americanus). The resting potential, action potential, and overshoot for these fibers were very similar in magnitude to the corresponding parameters for the nodes of Ranvier of frog medullated nerve fibers: ~70 mV, 118 mV, and 47 mV, respectively.

Reducing the sodium ion concentration in the solution to 25% of its normal level (replacing NaCl by choline chloride) led to a marked decrease in amplitude of the action potential. In this case, just as in the experiments on the squid giant axons, the decrease in amplitude of the spike at low values of $[Na]_0$ was much greater than would be expected for a sodium electrode. In some cases conduction was blocked when the value of $[Na]_0$ was reduced to 50% of its normal level.

Investigations have shown that the action potential is generated by calcium, and not sodium, ions in a few excitable structures: crustacean muscle fibers (Fatt and Katz, 1953; Fatt and Ginsborg, 1958; Liberman, 1963), the giant muscle fibers of the acorn barnacle (Hagiwara, Chichibu, and Naka, 1964), giant neurons of some gastropod mollusks (Gerasimov, Kostyuk, and Maiskii, 1965a,b), and smooth muscle fibers (Kuryama, Osa, and Toida, 1967).

Movement of Ions Through the

Membrane During Activity

Investigation by the Use of Labeled Atoms

The view that sodium and potassium ions play a role in action potential generation in nerve and muscle fibers, described above, has also been directly confirmed by investigations of the movement of ions through the membrane during generation of action potentials.

By the use of K^{42} and Na^{24}, Keynes (1951a) found that stimulation of the cuttlefish giant axon at a frequency of 100/sec increases both the ingoing and outgoing fluxes of these ions. The influx of K^+ ions was increased by 3.3 times, and the efflux by 9.1 times.

Taking the internal potassium concentration as 246 mM, Keynes calculated that when each impulse is conducted along the investigated nerve fibers, between 3.9×10^{-12} and 5.5×10^{-12} mole/cm^2 of K^+ ions leave the fiber. The membrane fluxes of Na^+ ions were very considerably increased during stimulation of the nerve fiber. The sodium influx was increased by 18 times and the efflux by 22 times.

These results showed conclusively that conduction of the nervous impulse is in fact associated with an increase in the permeability of the membrane to sodium and potassium ions. However, they did not answer the question of whether under these circumstances there is a resultant increase in the Na^+ ion concentration in the cytoplasm. Yet to prove the validity of the "sodium hypothesis," it was necessary to show that the number of sodium ions entering the axon during each impulse is sufficient to enable the membrane capacitance to be recharged to the full value of the action potential.

In order to answer this question, Keynes and Lewis (1951) used the method of activation analysis, by means of which Na^+ and K^+ can be determined in amounts as low as 0.3 μg and 3 μg respectively, with an accuracy of 2%. The nerve fibers tested were irradiated for a week with neutrons, so that definite numbers of labeled Na^{24} and K^{42} ions were formed in their axoplasm. The irradiated samples were then counted (β - and γ-rays) with the appropriate fibers, thus giving the concentrations of potassium and sodium in the axoplasm simultaneously. This investigation showed that during activity evoked by stimulation for 20 min at 200 sec^{-1} the sodium concentration in the nerve fiber rises by 3×10^{-12} mole/cm^2/pulse and the potassium ion concentration falls by 3.6×10^{-12} mole/cm^2/pulse. This corresponds to the passage of approximately 20,000 ions through 1 μ^2 of fiber surface. It was also found that during activity of the axon there is a very small increase in the concentration of Cl^- ions in its axoplasm ($\sim 0.6 \times 10^{-12}$ mole/cm^2/pulse).

Similar results were obtained on giant axons by Rothenberg (1950). In his experiments stimulation of the fiber for 30 min at 100 sec^{-1} led to an increase in the Na^+ ion concentration in the cytoplasm by 4.5×10^{-12} mole/cm^2/pulse.

If it is assumed that the capacitance of the giant axon membrane is approximately 1 μF/cm^2 (Curtis and Cole, 1938), a change in potential of 120 mV must require 1.2×10^{-7} coulomb, i.e., 1.2×10^{-12} mole of monovalent cations. This is approximately one-third of the number of Na^+ ions observed entering the axon experimentally. The difference is due to the fact that Na^+ ions enter the fiber not only during the ascending phase of the action potential, but also during its descending phase, in exchange for K^+ ions leaving the fiber (see Fig. 30).

The efflux of K^+ ions, as Shanes (1954) showed, increases by almost 3 times (from 3 to 9 pmole/cm^2/pulse) on cooling the axon from 24 to 6.1°C, i.e., by approximately the same factor as the descending phase of the action potential is prolonged.

Investigation of Ionic Currents by the Voltage-

Clamp Method

Modern interpretations of the laws governing movement of ions through a membrane are based principally on results obtained by the voltage-clamp method.

The essence of this method is measurement of the current flowing through a definite area of the membrane of a nerve fiber under conditions in which the membrane potential throughout this area is maintained automatically, by means of a feedback amplifier, at an assigned level. The first experiments in this direction were begun in 1947 by Marmont (1949) and Cole (1949), who developed a technique of introducing a long metal electrode inside the giant axon of a cuttlefish. By this means, when passing an electric current through the membrane they were able to avoid all complications due to spread of the current along the cable-like structure of the fiber (see pages 38 and 213).

Cole used only one electrode inserted into a fiber to measure the membrane potential and also to apply the current. However, polarization of the electrode arising during the passage of a strong current prevented measurements lasting longer than 1 msec. Despite these difficulties, Cole (1949) was able to show for the first time that depolarization of the membrane is accompanied by a powerful inward ionic current.

Important improvements to this technique were made by Hodgkin, Huxley, and Katz (1949, 1952).*   These workers suggested introducing two thin wire electrodes inside the giant axon and along its axis, one to record the membrane potential and the other to pass the current. The source of the current was the output of an amplifier which compared the measured membrane potential to a reference emf. In this way the membrane potential could be changed instantaneously by any desired amount and maintained stable at its new level.

Later, similar systems of "voltage clamp" were developed for other objects, including single nodes of Ranvier from frog medullated nerve fibers (Dodge and Frankenhaeuser, 1958; Dodge, 1963), crustacean nonmedullated nerve fibers (Julian, Moore, and Goldman, 1962), single nerve cells (Hagiwara and Saito, 1959), Purkinje fibers of the heart (Deck and Trautwein, 1964; and others), and skeletal muscle fibers (Adrian, Chandler, and Hodgkin, 1970).

---

*For a review of the history of these classical investigations, see Cole (1968).

In complete agreement with the theoretical predictions, these experiments showed that the net membrane current (I_m) generated in a uniform area of membrane in response to a potential change (V) consists of a capacitance current ($C\dot{V}$), resulting from a change in the ion density on the outer and inner surfaces of the membrane (the displacement current), and an ionic current (I_i), resulting from a movement of charged particles through the membrane:

$$I_m = C\frac{dV}{dt} + I_i \tag{14}$$

If the membrane potential is maintained at a constant level, $\dot{V}$ (and, consequently, $C\dot{V}$) will be equal to zero, so that the ionic current can be obtained directly from the experimental records of I_m.

It has been shown by various methods (see below) that the net ionic current (I_i) in the squid giant axon can be subdivided into three principal components: the initial sodium current (I_{Na}), the delayed potassium current (I_K), and the leak current (I_l).

A fourth component of the ionic current — the delayed "nonspecific current" (I_p), has also been found in medullated nerve fibers. This component is carried mainly by Na^+ ions, but it can perhaps also be carried by other ions, such as Ca^{++} (Frankenhaeuser, 1963a).

During changes of membrane potential, the sodium and potassium current (and I_p also in medullated fibers) undergoes complex changes connected with specific changes in ionic permeability of the membrane. By contrast, the leak conductance (g_l) during depolarization or hyperpolarization is virtually unchanged, so that the leak current (I_l) is usually taken as directly proportional to the change in membrane potential.* During hyperpolarization of the membrane, I_l is inward in direction, while during depolarization it is outward. Since an increase in the concentration of K^+ and Na^+ ions in the solution increases the leak current slightly (Frankenhaeuser and Huxley, 1964), it is suggested that both ions participate in the transport of this current. The contribution of Cl^- ions to the leak current in nerve fibers is evidently small.

---

*In the light of results published by Adelman and Taylor (1961) and by Goldman and Binstock (1969), however, this statement requires certain qualification.

For instance, Frankenhaeuser (1962b) replaced the Cl⁻ ions in Ringer's solution by less permeable anions, namely isethionate and methylsulfate, but this had no effect on the magnitude of the leak current of a node of Ranvier. According to Brinley and Mullins (1965), the transport number for chloride ions, i.e., the proportion of the charges carried by these ions, in squid giant axons is only 0.04.

By contrast with the membrane of nerve fibers, the membrane of muscle fibers has high chloride conductance. According to Hodgkin and Horowicz (1959b), at the resting potential the transport number of Cl⁻ ions is twice that of K⁺ ions.

To measure the leak current, usually the membrane potential is changed toward hyperpolarization.

In the experiment illustrated in Fig. 20a, the membrane of a node of Ranvier was hyperpolarized stepwise by 60 mV. Very brief surges of membrane current, hardly distinguishable on the record, due to charging of the membrane capacitance to the new level, appeared at the beginning and end of the step.

Immediately after the capacitance component the ionic component of the current — the leak current — appeared. It corresponded exactly in magnitude and direction to the current calculated by Ohm's law from the values of the potential change (V) and the resistance of the membrane (R). Throughout the period of hyperpolarization the strength of the leak current remained unchanged.

During depolarization of the membrane a leak current also was generated, but it was masked by the sodium and potassium currents due to specific changes in the permeability of the membrane to these ions.

Changes in the ionic current of the membrane of the node of Ranvier during depolarization by 60 mV are shown in Fig. 20b.

Initially in this case a short surge of capacitance current likewise is observed; it is followed by an outward leak current which is replaced very quickly by a wave of inward current connected with the movement of Na⁺ ions along the concentration gradient. The inward current rapidly reaches its maximum, after which it declines and changes into the outward wave of the potassium current.

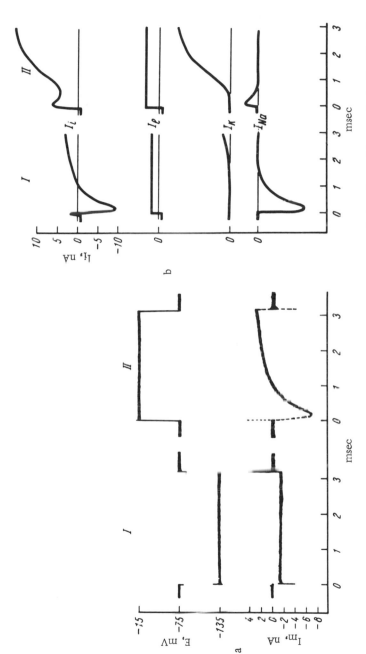

Fig. 20. Ionic currents in a single node of Ranvier: a) changes in net ionic current during clamped hyperpolarization (I) and depolarization (II) of the membrane, 20°C; b) analysis of the net ionic current into its components. I) membrane preliminary hyperpolarized to $E = -105$ mV, and then depolarized at the moment $t = 0$ to $E = 0$ mV; II) membrane preliminarly hyperpolarized to $E = -105$ mV, and then depolarized to $+67.5$ mV, I_{Na} was therefore outward. Abscissa, time in msec; ordinate, strength of ionic current (I_i) in nA. Explanation in text. Node of Ranvier, 20°C (Dodge, 1963).

If Na^+ ions are removed from the Ringer's solution surrounding the fiber, the sodium component of the net ionic current disappears, after which only the leak current and the specific potassium current, which are outward in direction, are recorded. The latter can be differentiated also by measuring the strength of the leak current during corresponding hyperpolarization of the membrane.

An example of this analysis of the net ionic current into its components I_l, I_K, and I_{Na} is illustrated in Fig. 20a,b. Examination of this figure shows clearly that during the first 0.5 msec after the beginning of depolarization of the membrane, when I_K is very small, the total ionic current is approximately equal to the sum $I_{Na} + I_l$. During prolonged depolarization of the membrane (in this case, when t > 2 msec), on the other hand, because of inactivation of the sodium permeability the sodium current falls sharply, and the ionic current is determined mainly by the sum $I_l + I_K$.

The magnitude and direction of the ionic membrane currents depend on the membrane potential.

The results of one of Dodge's (1963) experiments on a node of Ranvier, in which the membrane potential was changed stepwise, are illustrated in Fig. 21a. During very weak depolarization of the membrane (from -75 to -60 mV) only the leak current appeared. During stronger depolarization (to -45 mV), besides the outward leak current, a weak inward sodium current appeared. As the strength of depolarization increased, the inward current of Na^+ ions increased progressively and reached its maximum at $E = -15$ mV. A further increase in the strength of depolarization of the membrane caused a decrease in I_{Na}. At $E = 35$ mV the inward I_{Na} disappeared completely, and at $E = 60$ mV it changed its direction.

A similar relationship between the net ionic current and the membrane potential was also found in experiments on squid giant axons, on nonmedullated crustacean nerve fibers, and on mollusk giant nerve cells. The reason for the increase in the initial inward current on increasing the strength of depolarization of the membrane is clear.

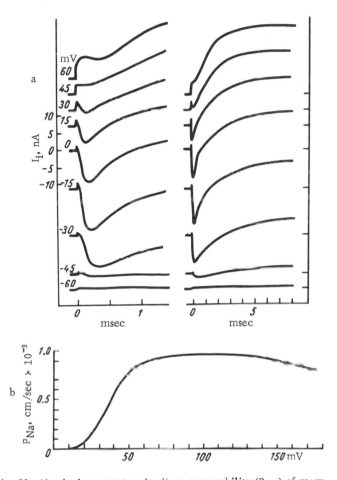

Fig. 21. Net ionic current and sodium permeability (P_{Na}) of membrane of node of Ranvier from amphibian nerve fiber as a function of change in membrane potential: a) kinetics of changes in net ionic current (I_i) during depolarization of the node membrane in a step from E = −75 mV to various levels of the membrane potential. The family of curves in the right half of the figure was obtained with a slower sweep than the family of curves on the left. Numbers in front of each curve indicate the value of E at which the voltage clamp took place, 22°C (Dodge, 1963); b) graph showing how P_{Na} (ordinate), measured at the time of the I_{Na} peak, varies with the clamped voltage (abscissa) (Dodge and Frankenhaeuser, 1959).

The sodium current depends on the permeability of the membrane to these ions and on the electrochemical potential, i.e., the difference between the membrane potential E and the sodium equilibrium potential E_{Na}. With an increase in the strength of depolarization of the membrane (an increase in E), P_{Na} rises while $E - E_{Na}$ falls. An increase in the potential inside the membrane to -15 mV causes a sharp increase in sodium permeability (Fig. 21b), so that, despite the decrease in the electrochemical potential, the peak value of I_{Na} is increased. With a further increase in the strength of depolarization, however, the rise in P_{Na} becomes smaller, and at positive values of E it disappears altogether (the P_{Na} curve reaches the plateau level, see Fig. 21b). As a result of this, I_{Na} weakens, passes through zero, and then becomes reversed.

Unlike the sodium current, the delayed potassium current I_K, measured at the plateau level (at $t = \infty$), rises steadily with an increase in the strength of depolarization.

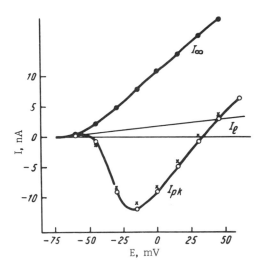

Fig. 22. Peak value of early (I_{pk}) and stationary value of late (I_∞) ionic current as functions of membrane potential. Abscissa: potential inside membrane; ordinate: ionic current I, in nA. I_l) Leak current. Graph plotted from results of experiment illustrated in Fig. 16b (Dodge, 1963).

The peak value (I_{pk}) of the initial ionic current and the value
of the outgoing ionic current measured at the plateau level (t = ∞)
are shown in Fig. 22 as functions of the membrane potential. The
leak current I_l is also shown as a function of the membrane potential in the same figure.

Evidence for Subdivision of the Ionic Currents

Let us now consider the facts concerning the validity of Hodgkin and Huxley's suggestion that the net ionic current is subdivided
into sodium and potassium components.

The Sodium Current

1. It was pointed out above that a decrease in the external Na^+
ion concentration in the solution leads to a reduction in the inward ionic current, while if Na^+ is completely removed from the
medium this current disappears completely. The potassium current is virtually unchanged under these conditions.

It was shown later that an effect similar to that produced by
removal of Na^+ from the medium can be obtained by application
of tetrodotoxin to the nerve fiber. Tetrodotoxin is a water-soluble
paralytic poison found in the tissues of the Japanese puffer fish
Spheroides rubrides and the tissues of the Californian salamander
(Taricha torosa). In very low concentrations it inhibits action potential generation in nerve and muscle fibers (see the survey by
Kao, 1966; see also Khodorov and Belyaev, 1967, and Khodorov and
Vornovitskii, 1967). Investigations by the voltage-clamp level on
giant axons of the lobster (Narahasi, Moore, and Scott, 1964), and
squid (Nakamura, Nakajama, and Grundfest, 1965; Takata, Moore,
Kao, and Fuhrman, 1966), and on single nodes of Ranvier (Hille,
1966, 1967a) showed that tetrodotoxin selectively blocks only the
initial (sodium) ionic current and has no effect on the late outward current associated with the transport of potassium ions (Fig.
23, I). This is evidence of the relative independence of the initial
and late components of the net ionic current.

2. Hodgkin and Huxley (1952a) showed in experiments on squid
giant axons that the potential at which the initial ionic current re-
verses its direction (the equilibrium potential for the initial cur-

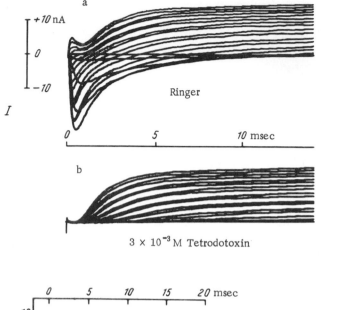

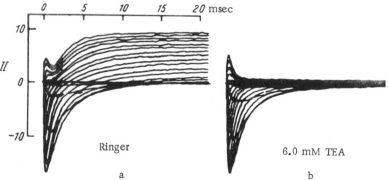

Fig. 23. Effect of tetrodotoxin and tetraethylammonium on ionic currents through the membrane of the node of Ranvier. Voltage clamped at different levels between -67 and $+67.5$ mV. Leak currents automatically deducted. I: a) Node in Ringer's solution; b) in solution containing 3×10^{-7} M tetrodotoxin (Hille, 1966). II: Node in Ringer's solution (a) and in 6 mM tetraethylammonium solution (b) (Hille, 1967b).

rent) is very close in value to the sodium equilibrium potential calculated by Nernst's equation.

It follows from this equation (13) that with changes in the external Na^+ ion concentration, the value of E_{Na} must change in the following manner:

$$\Delta E_{Na} = E'_{Na} \text{ (experimental solution)} - E''_{Na} \text{ (seawater)} = \frac{RT}{F} \ln \frac{[Na]'_0}{[Na]''_0}$$

By varying $[Na]'_0/[Na]''_0$ between 0.1 and 0.3, these workers obtained good agreement between the experimentally observed and theoretically expected changes in ΔE_{Na}.

Similar investigations were carried out by Dodge and Frankenhaeuser (1958) on a single node of Ranvier.

A decrease in the external Na^+ concentration from 112 to 39.9 mM lowered the equilibrium potential for the inward ionic current by 18-24 mV. The equation predicted that under these conditions the values of E_{Na} must fall by 25 mV. The small divergence between the experimental and theoretical values was attributed to unavoidable errors of measurement during work by their voltage-clamp method. They estimated these errors to be 1-7 mV.

Frankenhaeuser and Moore (1963) later found that the initial increase in ionic permeability of the membrane during depolarization of the node of Ranvier is not absolutely specific for Na^+ ions. The potassium permeability is simultaneously increased, and the ratio between the coefficients of sodium and potassium permeability at the height of the peak of the initial ionic current is about 20:1.

This ratio is so great that the initial current can be regarded virtually as a current carried by Na^+ ions. However, the fact that a certain proportion of it is carried by K^+ can give a partial explanation of why the change in equilibrium potential for this initial current associated with a decrease in $[Na]_0$ is a little smaller than the theoretically calculated value of ΔE_{Na}.

Similar results were soon obtained on squid giant axons.

Chandler and Meves (1965) perfused giant axons after preliminary extrusion of their cytoplasm with solutions containing different concentrations of Na^+ ions. Reducing the sodium concentration in the perfusion solution led to a decrease in the initial component of the ionic current obtained during fixation of the membrane potential at a level higher than E_{Na}. However, this component did not disappear even after complete removal of Na^+ from the internal solution. Analysis of these results led Chandler and Meves to conclude that in the absence of Na^+ inside the axon, K^+ ions can pass through the sodium channels. However, the maximum permeability of these channels for K^+ is only one-twelfth of their permeability to Na^+ ions.

These workers give the following series of relative permeabilities of the sodium channels for different ions:

$$Li : Na : K : Rb : Cs = 1.1 : 1 : \frac{1}{12} : \frac{1}{40} : \frac{1}{61}.$$

3. The direct verification of Hodgkin and Huxley's hypotheses on the role of Na^+ ions in the genesis of the initial ionic current was undertaken by Mullins, Adelman, and Sjodin (1962), who combined the voltage-clamp method with the method of labeled atoms.

The giant axon was stimulated for a long time before the experiment began in a solution containing Na^{24}. This resulted in a high concentration of radioactive sodium in its axoplasm. The membrane potential of the axon was then clamped at a level higher than E_{Na} (+10 mV) so that the initial current was outward in direction. When the initial current had reached its maximum, 0.5 msec after the beginning of clamping the potential returned to its initial level and then increased again up to 100 mV. The nerve fiber was thus exposed to the repetitive action (10-20/sec) of short clamping pulses of emf.

By integrating the initial ionic current with respect to time, these workers obtained values for the quantity of electricity carried through the membrane during each pulse. If the axon was kept in sodium-free medium (seawater containing choline), each pulse corresponded to 1.2 $\mu C/cm^2$. The measured outflow of Na^{24} under these conditions was 1.4 $\mu C/cm^2$. The transport number of the sodium ions (T_{Na}) was therefore approximately 1.2.

Even clearer results were obtained by Atwater, Bezanilla, and Rojas (1969) by the use of similar experimental methods on internally perfused axons of the Chilean squid Dosidicus gigas. By means of internal perfusion it was possible to measure the inward flow of sodium ions (Na^{22} was added to the external solution) and to use depolarizing pulses of different amplitudes. During all the membrane potential steps the inward ionic current measured electrically was roughly identical with the inward current recorded directly. In these experiments the value of T_{Na} was 0.92.

By using the same method, but varying the duration of the depolarizing pulses, Bezanilla et al. (1970) for the first time were able to study the time course of changes in sodium permeability directly. In these experiments also, the electrical measurements agreed well with the direct measurements of the sodium flux.

The Potassium Current

The delayed outward current is generally identified as a potassium current on the basis of the following facts:

1. In experiments on nerve fibers of Sepia officinalis, Hodgkin and Huxley (1953) showed that during prolonged depolarization of the membrane, equivalent relationships are found between the number of K^+ ions (K^{42} was used) leaving the axon and the quantity of electricity passing through the membrane in the same period of time.

Brinley and Mullins (1965) injected K^{42} into the squid giant axon for a length of 10-20 mm and passed an outward current through an electrode lying inside the axon and centered in the zone of the injection. The transport number (T_K) was 1.0 ± 0.3. The ratio between T_K and T_{Na} during prolonged depolarization of the membrane in these experiments was 20:1 or more. $T_{Cl} \sim 0.04$ and $T_{Na} \sim 0.05$.

The results obtained by Mullins, Adelman, and Sjodin in their investigations cited above (1962) were less precise.

The outward current of potassium ions from an axon, preliminarily enriched with K^{42}, into seawater was only 50% of the charges transported in cases in which the membrane was kept for 5 msec at a potential equal to E_{Na}, i.e., under conditions in which the initial sodium current was absent. Only in axons with a low membrane potential (of the order of −45 mV) was T_K approximately unity.

In giant squid axons Hodgkin and Huxley (1952a) did not obtain complete agreement between E_K, measured by the voltage-clamp method, and E_K calculated by Nernst's equation for different values of $[K]_0$.

By contrast, Frankenhaeuser (1962a) obtained good agreement in experiments on single nodes of Ranvier from nerve fibers of Xenopus laevis between the values of I_K measured experimentally with different K^+ ion concentrations in the medium and the values calculated by means of the Hodgkin and Huxley equations.

2. Just as tetrodotoxin selectively blocks the sodium conductance of the membrane without changing the value of P_K, the quaternary ammonium base tetraethylammonium has a selective effect on the potassium permeability channels.

Considerable diminution, down to zero, of the outward I_K under the influence of tetraethylammonium was found in experiments using the voltage-clamp method on giant axons (Tasaki and Hagiwara, 1957; Armstrong and Binstock, 1965), on nerve cells of the marine mollusk Ochnidium verruculatum (Hagiwara and Saito, 1959), and the Atlantic puffer Spheroides maculatus (Nakajima, 1966), as well as on single nodes of Ranvier from nerve fibers of Xenopus laevis (Koppenhöfer, 1966, 1967; Hille, 1967a,b). The inward sodium current under these conditions remained virtually unchanged, or was very slightly reduced (Fig. 23b). Cesium ions, injected inside the giant axon, have a similar effect on the outward potassium current. On adding Cs^+ to the liquid perfusing an axon, Chandler and Meves (1965) and Adelman and Senft (1966) found a marked decrease in strength of the late outward ionic current. Using K^{42}, Sjodin (1966) showed that this effect of Cs^+ ions is in fact due to a sharp decrease (down to 7–22% of the initial value) in the loss of K^+ ions from the nerve fiber.

All these results confirm the validity of Hodgkin and Huxley's basic assumption regarding the relative independence of the sodium and potassium currents in nerve fibers. Each of these currents evidently has its own channels which are characterized by high selective permeability to ions of a particular type (see page 269).

Changes in Ionic Permeability

Subdivision of the net ionic current into its components has allowed a number of investigators to analyze the kinetics of changes in the sodium and potassium permeability of the membrane during changes in potential (Hodgkin and Huxley, 1952b; Frankenhaeuser, 1960–1963; Dodge, 1963).

An important method which has permitted investigations of this type to be undertaken is the two-step change in membrane poten-

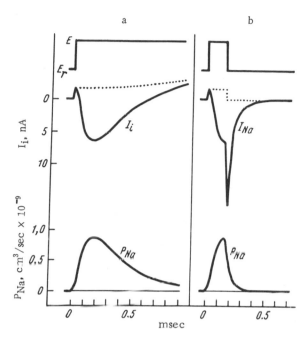

Fig. 24. Reversible nature of activation of sodium permeability: a) curve of net ionic current I_i and P_{Na}
during prolonged depolarization to $E = 0$ mV; b) curves
of I_{Na} and P_{Na}, associated with sudden repolarization of
the membrane at the time when P_{Na} reaches its peak value.
Changes in $I_l + I_K$ are shown by dots on the curve. Compare the rapid recovery of P_{Na} during repolarization of
the membrane (b) with the slow inactivation of P_{Na} during
prolonged depolarization (a) (Dodge, 1963).

tial. Initially the membrane was depolarized to a certain level,
and then after different intervals of time, i.e., in different phases
of the changes in ionic permeability, the membrane potential was
returned in one step to its initial value or was established at a
new level (Fig. 24). In this way the dependence of the ionic current on the electrochemical potential $(E - E_{Na}$ or $E - E_K)$ could
be studied.

In experiments along these lines it was shown that the permeability of the membrane to Na^+ and K^+ ions changes during
depolarization and hyperpolarization of the membrane at a finite
speed. This means that during an instantaneous change in potential the permeability remains the same as it was before the change,

and it only begins to change later, after a fraction of a milli-second.

During stepwise depolarization of the membrane the sodium permeability rises in an S-shaped curve, to reach a certain maximum, after which it begins to fall, and it ultimately returns to its initial low level. Within certain limits, the greater the shift in membrane potential toward depolarization, the higher the rate at which the sodium permeability rises and the higher the maximum which it reaches.

The potassium permeability also rises in an S-shaped curve during depolarization of the membrane, but the rate of its increase, for a given shift of potential, is slower than the rate of increase in sodium permeability. Another difference is that during continued depolarization the potassium permeability remains for a long time at a more or less stationary level.

Only during depolarization which continues for hundreds of milliseconds does the potassium permeability also begin to fall, ultimately to become stabilized at a certain stationary level (see page 93).

To explain the diphasic sodium permeability changes Hodgkin and Huxley postulated that membrane depolarization induces two simultaneous but opposite processes: rapid activation of the sodium channels and their slow inactivation. The inactivation process does not simply lower the sodium conductance to the initial low value, but it brings the system regulating the passive transport of these ions through the membrane into a refractory state. As a result of this the second depolarizing stimulus, applied against the background of the first stimulus or immediately after its end, is unable to induce a fresh increase in sodium permeability and in the sodium current to its former level. The stronger and the more prolonged the preliminary depolarization of the membrane, the more marked the inactivation of the sodium system. Inactivation develops slowly (Fig. 25a), and it is abolished equally slowly after restoration of the initial level of the membrane potential (Fig. 25b).

The decrease in potassium permeability during very prolonged depolarization indicates that, besides sodium inactivation, potassium inactivation also takes place (see page 93). Potassium inactivation differs from sodium in its very high time constant (ap-

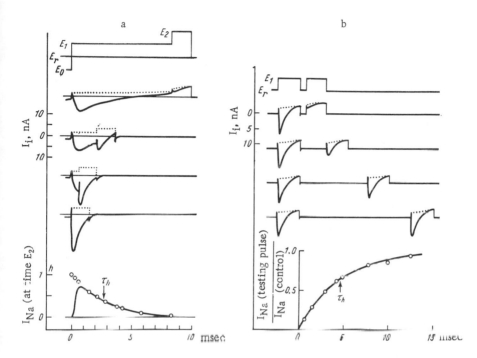

Fig. 25. Kinetics of inactivation: a) simultaneous and independent development of activation and inactivation. Before each measurement the membrane was hyperpolarized to $E_0 = -105$ mV for 40 msec. Immediately after this, depolarization was carried out in a single step to $E_1 = -45$ mV, for various lengths of time, after which depolarization was strengthened by a further step to $E_2 = -15$ mV. Broken lines indicate changes in $I_l + I_K$. The ratio between the peak value of I_{Na} during the second step E_r and I_{Na}, obtained during instantaneous depolarization of the membrane from E_0 to E_r (curve below), is taken as the index of magnitude h. A change in h during depolarization is shown on the lower diagram. $\tau_h = 2.7$ msec. This process clearly arises soon after the beginning of depolarization E_1; b) reactivation of sodium permeability after end of depolarizing pulse E_1. With an increase in the length of the interval between the facilitating and testing pulses, the value of I_{Na} to the second pulse increases and returns to the initial control value. Recovery curve shown below. Its time constant $\tau_h - 4.6$ msec (Dodge, 1963).

proximately three orders of magnitude larger) and also in the fact that P_K never falls to zero as the result of potassium inactivation.

Quantitative Description of

Ionic Currents

The results obtained by the voltage-clamp method have enabled the ionic currents through a membrane to be described quantitatively, with the subsequent formulation of a mathematical theory of the nervous impulse. This theory was originally developed for the membrane of the squid giant axon (Hodgkin and Huxley, 1952c; Huxley, 1959; Hodgkin, 1964b), and it was later applied and modified for amphibian medullated nerve fibers (Frankenhaeuser and Huxley, 1964; Dodge, 1963).

A Mathematical Model of the Membrane of

the Squid Giant Axon

It was pointed out above that the force determining movement of ions through a membrane is the electrochemical potential, i.e., the potential difference between the membrane potential (E) and the equilibrium potential for that particular ion.

The electrochemical potential for Na^+ ions is equal to $E - E_{Na}$. It can be expressed also in relative units if, instead of the absolute values of E and E_{Na}, the values V and V_{Na}, which show by how much E and E_{Na} differ from the resting potential (E_r), are substituted:

$$V = E - E_r; \quad V_{Na} = E_{Na} - E_r;$$

whence

$$E - E_{Na} = V - V_{Na}.$$

The electrochemical potential for K^+ ions can also be expressed as the difference

$$E - E_K = V - V_K.$$

Finally, the electrochemical potential for ions participating in the leak current is given by

$$E - E_l = V - V_l.$$

By calculations and direct measurements the following values have been obtained for equilibrium potentials of squid giant axons: $E_{Na} = +55$ mV; $E_K = -72$ mV; $E_l = -50$ mV.

Experiment showed that at any instant the sodium, potassium, and leak currents in giant axons are linear functions of the corresponding electrochemical potentials:

$$I_{Na} = g_{Na}(V - V_{Na})$$ (15)

$$I_K = g_K(V - V_K)$$ (16)

$$I_l = g_l(V - V_l),$$ (17)

where g_{Na}, g_K, and g_l represent the sodium, potassium, and leak conductances respectively.

This relationship is linear because the conductance of the membrane to Na^+ and K^+ ions is independent of the direction of the ionic current: they are equal no matter whether the ions flow with or against the concentration gradient. This means that the sodium and potassium conductances of the membrane in the giant axon are directly proportional to the permeability of the membrane to these ions. We shall see below that in medullated nerve fibers, whose membrane possesses rectifying properties, more complex relationships hold well between ionic conductance and permeability calculated from constant field equations (Fig. 36).

To describe the relationship between potassium conductance, on the one hand, and membrane potential and time on the other hand, the following equations have been suggested:

$$g_K = \bar{g}_K n^4$$ (18)*

$$\frac{dn}{dt} = \alpha_n(1 - n) - \beta_n \cdot n.$$ (19)

In these equations $\bar{g}_K$ represents the "maximal potassium conductance" or the constant of potassium conductance; n the variable of the process of activation of the potassium channels; α_n and β_n the rate constants, the values of which depend on the potential (Fig.

---

* Cole and Moore (1960) suggested that n should be raised not to the 4th, but to the 25th power to describe the very considerable delay in the rise of g_K during an increase in V after preliminary strong hyperpolarization of the membrane.

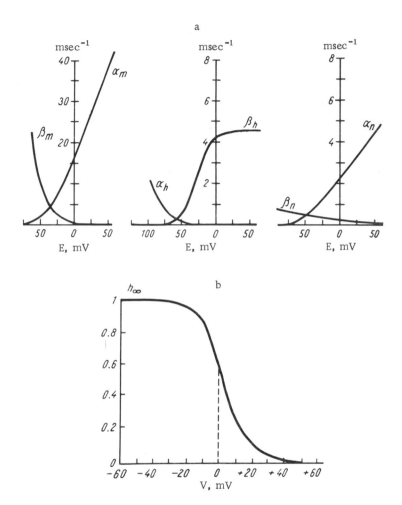

Fig. 26. Rate constants α_m, β_m, α_h, β_h, α_n, and β_n and stationary values of h_∞ as functions of membrane potential of the squid giant axon: a) abscissa, membrane potential E in millivolts; ordinate, rate constants in msec^{-1}; b) abscissa, shifts in membrane potential (V) toward hyperpolarization (with minus sign) and depolarization (with plus sign). Initial value V = 0; ordinate, magnitude of h_∞ characterizing proportion of sodium channels accessible for activation during depolarization (Hodgkin and Huxley, 1952c).

26a), the temperature, and the Ca^{++} concentration in the medium, and are independent of time.

During depolarization of the membrane, n rises from near zero to a certain stationary level n_∞ (during strong depolarization, $n_\infty \to 1$), along an exponential curve with a time constant of

$$\tau_n = \frac{1}{\alpha_n + \beta_n}. \tag{20}$$

The value of g_K rises correspondingly along an S-shaped curve, as is observed experimentally.

Upon return of the membrane potential to its initial level, α_n decreases to a value smaller than β_n, so that g_K falls along an exponential curve.

In an attempt to give a physical interpretation of these empirically chosen equations, Hodgkin (1964) put forward the following suggestion.

It can be assumed that the K^+ ion passes through the corresponding channel in the membrane only if four charged "activating" particles reach the region of this channel which is under the influence of the electric field. If n is the probability that a single particle is in the right position, n^4 is the probability that all four particles will be in the correct position. From this point of view, the term α_n characterizes the velocity of movement of the particle toward the potassium channel, while β_n characterizes the velocity of its movement in the opposite direction.

Changes in sodium conductance during depolarization are determined by two independent* processes, running in opposite directions — activation and inactivation:

$$g_{Na} = \bar{g}_{Na} m^3 \cdot h, \tag{21}$$

where $\bar{g}_{Na}$ is the maximum sodium conductance; m is the variable

---

*The view that these processes are independent is disputed by Hoyt (1963, 1968). In the model which she proposes, g_{Na} depends on one variable (V) only, but V is determined by two coupled processes. A similar model was suggested by Gul'ko (1968). The predictions of Hoyt's model relative to most phenomena are strikingly similar with the corresponding predictions of the Hodgkin—Huxley model. Differences are found only in the dependence of inactivation effects on the testing potential (see Hoyt and Adelman, 1970).

of the process of activation of the sodium channel, and h the variable of the inactivation process.

During depolarization of the membrane m rises while h falls. The changes in m and h with time are determined by the rate constants α_m, β_m, α_h, and β_h:

$$\frac{dm}{dt} = \alpha_m (1 - m) - \beta_m \cdot m \tag{22}$$

$$\frac{dh}{dt} = \alpha_h (1 - h) - \beta_h \cdot h. \tag{23}$$

Just like α_h and β_h, the above rate constants are dependent on membrane potential (Fig. 26a) and temperature, but are independent of time. The value of Q_{10} for all these rate constants is approximately 3.

By analogy with the interpretation which was given when the relationship between g_K and the potential and time was examined, Hodgkin postulated that the Na^+ ion can pass through the corresponding channel of the membrane only when three charged "activating" particles are present in the region of this channel which is under the influence of the electric field. If m is the probability that a single activating particle is in the corresponding position, m^3 is the probability that all three particles have reached there simultaneously, so that the sodium conductance is proportional to the value of m^3. However, besides activating particles, particles responsible for inactivation of sodium conductance also move under the influence of the electric field into the region of the sodium channel. The probability of one inactivating particle arriving in this part of the membrane can be expressed by the term $1 - h$. In the same way, the probability that no inactivating (blocking) particle will yet have reached the sodium channel at the time when three activating particles have arrived there will be given by the term h. The sodium conductance is thus proportional to $m^3 \cdot h$.

With this interpretation of the events the rate constants α_m and β_h characterize the velocity of movements of the activating and inactivating particles into the region of the sodium channel during depolarization of the membrane, while the rate constants β_m and α_h are determined by the velocity of movement of these particles in the opposite direction during restoration of the membrane potential.

With the stationary value of the membrane potential (i.e., when t = ∞), the value of h (h_∞) is determined by the equation:

$$h_\infty = \frac{\alpha_h}{\alpha_h + \beta_h} \tag{24}$$

The curve of h_∞ versus V shows the dependence of steady-state sodium inactivation on the level of the membrane potential. In the case examined in Fig. 26b, for a resting potential V = 0, the value of h_∞ is 0.6. This means that 60% of the maximum sodium conductance can be activated by depolarization of the fiber membrane. Correspondingly, 40% of the sodium channels are inactivated.

During prolonged (~50 msec) hyperpolarization of the membrane by 50 mV this inactivation is abolished and h_∞ becomes equal to unity. During prolonged depolarization, on the other hand, the fraction of inactivated channels is increased, and at V = 50 mV, h_∞ is approximately zero.

The curve of h_∞ versus V can be described by the empirical equation

$$h_\infty = \frac{1}{1 + \exp\dfrac{V - V_h}{7}}, \tag{25}$$

where V_h is the change in potential sufficient to reduce h_∞ from its initial value at the resting potential to 0.5. In squid giant axons, V_h was found to vary from 1.5 to 3.5 mV.

The complete system of equations simulating the behavior of the squid giant axon membrane is as follows:

$$I_m = C\frac{dV}{dt} + \bar{g}_{Na}m^3h\left(V - V_{Na}\right) + \bar{g}_K n^4\left(V - V_K\right) + g_l\left(V - V_l\right) \tag{26}$$

$$\frac{dm}{dt} = \alpha_m\left(1 - m\right) - \beta_m \cdot m \tag{22}$$

$$\frac{dh}{dt} = \alpha_h\left(1 - h\right) - \beta_h \cdot h \tag{23}$$

$$\frac{dn}{dt} = \alpha_n\left(1 - n\right) - \beta_n \cdot n \tag{19}$$

$$\alpha_m = 0.1\,\frac{V - 25}{1 - \exp\dfrac{25 - V}{10}} \tag{27}$$

$$\beta_m = 4 \exp\left(-\frac{V}{18}\right) \tag{28}$$

$$\alpha_h = 0.07 \exp\left(-\frac{V}{20}\right) \tag{29}$$

$$\beta_h = \frac{1}{1 + \exp\dfrac{30 - V}{10}} \tag{30}$$

$$\alpha_n = 0.01 \frac{V - 10}{1 - \exp\left[(10 - V)/10\right]} \tag{31}$$

$$\beta_n = 0.125 \exp\left(-\frac{V}{80}\right) \tag{32}$$

$$C_m = 1 \quad \mu\text{F/cm}^2, \quad \bar{g}_{\text{Na}} = 120 \quad \text{mmho/cm}^2, \quad \bar{g}_{\text{K}} = 36 \quad \text{mmho/cm}^2$$

$$g_l = 0.3 \quad \text{mmho/cm}^2$$

$$E_r = -60 \quad \text{mV}, \quad V_{\text{Na}} = 115 \quad \text{mV}, \quad V_{\text{K}} = -12 \quad \text{mV}, \quad V_l = 10 \quad \text{mV},$$

$$E_{\text{Na}} = 55 \quad \text{mV}, \quad E_{\text{K}} = -72 \quad \text{mV}, \quad E_l = -50 \quad \text{mV}.$$

Temperature, 6.3°C

Solution of these equations, at first manually (Hodgkin and Huxley, 1952), and later by means of a digital computer (Cole, Antoziewicz, and Rabinowitz, 1955; Cole, 1958; Huxley, 1959) showed that the suggested mathematical model reproduces all the principal phenomena of electrical activity of the giant axon satisfactorily, namely: the membrane and propagated action potentials, changes in impedance during the spike and after-potentials, the net sodium and potassium fluxes into and out of the fiber during the action potential, restoration of excitability in the relative refractory stage, anode-break excitation, the oscillatory responses of the membrane in a medium deficient in calcium, changes in action potential under the influence of cooling and heating of the fiber, abolition of the action potential by anodic pulses, hyperpolarization responses, repetitive discharges of spikes to a steady current, etc.

These findings were of great importance as confirmation of the validity of the Hodgkin – Huxley hypothesis.

If, in fact, the subdivision of the net ionic current into its components were incorrect, or if the rate constants of sodium activation or inactivation depended not only on membrane potential but also, for example, on its first derivative with respect to time, the solution of these equations would not have yielded an action potential similar to those observed experimentally.

It must be emphasized that Hodgkin and Huxley in no way regarded their suggested equations as final. They stress that these equations describe only rapid events taking place in the membrane during and immediately after the action potential, but they are not adequate for the analysis of changes in permeability and potentials lasting for periods measured in seconds or minutes (see pages 210, 211).

"Hodgkin and I," remarked Huxley in his Nobel speech, "consider that these equations must be regarded as a first approximation which in many respects requires further clarification and development during the search for the actual mechanism of the changes in ionic permeability at the molecular level." (Huxley, 1964, p. 1159.)

Medullated Nerve Fibers

The basic principles governing changes in ionic currents during depolarization and hyperpolarization of the node of Ranvier proved to be fundamentally similar to those found in experiments on squid giant axons. However, some significant differences were discovered between them.

1. In squid giant axons, as was stated above, I_{Na} at the initial moment of the change in membrane potential is a linear function of the electrochemical potentials, so that the instantaneous value of g_{Na} is always constant and is independent of the direction of flow of the sodium current associated with that particular displacement of membrane potential.

Other relationships between V versus V_{Na} and I_{Na} have been found in medullated nerve fibers.

The work of Dodge and Frankenhaeuser (1959) showed that during strong depolarization, when the sodium current becomes outward in direction, sodium conductance and, correspondingly, I_{Na} increase by a somewhat smaller amount than would be expected if g_{Na} were not dependent on the direction of the sodium current.

To explain this fact, these workers made use of data in the literature on the properties of artificial semipermeable membranes. Such membranes show characteristic rectification, as a result of which, if the electrochemical potential is the same, the ionic currents are greater in magnitude if the ions move through the membrane from the solution in which their concentration is high into the solution in which it is low, than if the ions moved in the opposite direction.*

This is explained by the fact that concentration of a given ion in a membrane and, consequently, its conductance is greater in the first case than in the second (Goldman, 1943; Katz, 1949; Teorell, 1949a,b). To find out whether the experimentally observed reduction in the increase in outgoing I_{Na} at high values of V can be explained in this way, Dodge and Frankenhaeuser (1959) used the constant field equation (5) suggested by Hodgkin and Katz (1949):

$$I_{Na} = P_{Na} \frac{F^2 E}{RT} \cdot [Na]_0 \cdot \frac{\exp\{(E - E_{Na})\,F/RT\} - 1}{\exp\{EF/RT\} - 1}. \qquad (33)^\dagger$$

Equation (33) was transformed, to give

$$P_{Na} = \frac{I_{Na} RT}{F^2 E} \cdot \frac{1}{[Na]_0 \exp} \frac{\exp\{EF/RT\} - 1}{\{(E - E_{Na})\,F/RT\} - 1}. \qquad (33a)$$

The values of P_{Na} at the peak of the inward sodium current were calculated by this equation, and the curve showing how these peak values of P_{Na} vary with the change in membrane potential V, examined above, was plotted (Fig. 21b).

This curve, as Fig. 21b shows, has an S-shaped rise, it reaches a plateau at V = 70-80 mV, and then with a further increase in V, it begins to fall slightly.

---

* This rectification must not be confused with the "delayed rectification" developing during prolonged depolarization of an excitable membrane and caused by an increase in its potassium permeability.

†In this equation the internal Na^+ ion concentration is described as follows:

$$[Na]_i = [Na]_0 \exp\left(-E_{Na}\,F/RT\right) \quad [\text{cf. Eq. (5)}].$$

Measurement of the momentary values of I_{Na} at different values of V showed that P_{Na}, unlike g_{Na}, is independent of the magnitude of the membrane potential at which this measurement is made.

It was later discovered that P_{Na} is independent of the Na^+ ion concentration in the medium: reducing this concentration to 37% of its initial value did not change P_{Na}.

On the basis of these results, the experimentally determined and theoretically calculated [Eq. (33)] relationships between the net ionic current and membrane potential were compared.

Agreement between the values predicted by theory and obtained experimentally proved to be perfectly satisfactory (Fig. 27).

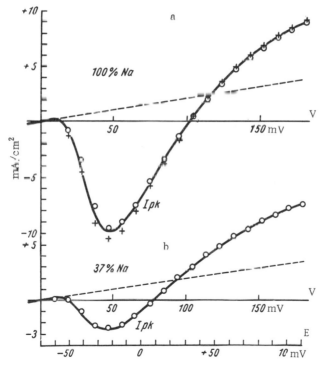

Fig. 27. Peak value (I_{pk}) of initial ionic current as a function of change in potential (V) toward depolarization: a) with normal Na^+ ion concentration in solution; b) with Na^+ ion concentration reduced to 37% of initial level. Broken line indicates leak current. Results obtained with single node of Ranvier. Smooth lines represent solution of Eq. (33) (Dodge and Frankenhaeuser, 1959).

To explain the fact that the membrane of the squid giant axon does not exhibit rectification, Hodgkin (see Frankenhaeuser, 1960) postulated that the electric field of this membrane is modified by the existence of fixed charges on one side: positive charges on the outside or negative on the inside (see also Bennett, 1967). If the potential step determined by these charges is equal in absolute magnitude to the sodium equilibrium potential, the Na^+ ion concentration inside the membrane must be equal regardless of whether the sodium current is inward or outward in direction. In that case the relationships between g_{Na} and P_{Na} become directly proportional.

2. A study of the kinetics of changes in P_{Na} led Frankenhaeuser to conclude that the dependence of P_{Na} on V and time can be better described quantitatively if the variable of the process of activation m is taken not as the third, as in the case of the squid giant axon model, but as the second power:

$$P_{Na} = \bar{P}_{Na} m^2 h, \tag{34}$$

where $\bar{P}_{Na}$ is the sodium permeability constant or the "maximum sodium permeability," and has the same meaning as in the Hodgkin–Huxley equations.

3. The membrane of the Ranvier node exhibited rectifying properties also in relation to the potassium current (Frankenhaeuser, 1962b), so that the term g_K in the corresponding equations describing I_K as a function of the potassium electrochemical potential $(V - V_K)$ was replaced by the coefficient of potassium permeability P_K:

$$P_K = I_K \frac{RT}{F^2 E} \frac{1 - \exp\{EF/RT\}}{[K]_0 - [K]_i \cdot \exp\{EF/RT\}}. \tag{35}$$

In agreement with the constant field theory, when the internal and external K^+ ion concentrations were equal, rectification of the potassium current disappeared.

4. Description of the dependence of P_K on V and time showed that the variable of activation of potassium permeability n must be taken, not to the fourth power, but to the second:

$$P_K = \bar{P}_K n^2 k, \tag{36}$$

where $\bar{P}_K$ is the constant of potassium permeability, or the "maximum potassium permeability," and k the variable of the process of potassium inactivation with the same significance as the variable h of sodium inactivation.

Potassium inactivation is found only during very prolonged depolarization of the membrane (>1 sec) and, consequently, it does not participate in action potential generation. Changes in k with time can be described by the equation

$$\frac{dk}{dt} = \alpha_K (1 - k) - \beta_K k, \tag{37}$$

where α_k and β_k are the rate constants. Their values are not completely determined at different values of V, but they are small enough to be disregarded during analysis of responses lasting less than 200 msec (Frankenhaeuser, 1963a; see also Moore, 1967).*

5. Besides the delayed potassium current in medullated nerve fibers, Frankenhacuser (1963a) also found a delayed "nonspecific" current I_p, transported by Na^+ ions and also, perhaps, by Ca^{++} ions (an increase in the Ca^{++} concentration strengthens I_p).

The nonspecific current I_p was detected when the ionic current arising immediately after the end of a prolonged depolarizing stimulus was measured. It is inward in direction.

The variable of the process of activation of the nonspecific current p obeys the same dependence on V and t as the variables m and n:

$$P_p = \bar{P}_p p^2, \tag{38}$$

$$\frac{dp}{dt} = \alpha_p (1 - p) - \beta_p \cdot p, \tag{39}$$

where α_p and β_p are the rate constants, dependent on V but independent of time. The values of α_p and β_p are small compared with the corresponding values of α_m and β_m at all values of V, so that the nonspecific current is found during depolarization only in the period when P_K of the membrane is increased.

---

* Ehrenstein and Gilbert (1966) conclude from their findings that the time constant of potassium inactivation in the giant axon at 9°C is approximately 11 sec.

6. The rate constants β_n and β_h were found to be dependent to some extent on time (on the "prehistory" of the nerve); preliminary depolarization of the membrane increases β_h and β_n (Frankenhaeuser, 1963a; Koppenhöfer, 1967).

However, when Frankenhaeuser and Huxley (1964) constructed their mathematical model of the membrane of the Ranvier node, they disregarded these facts.

Investigation of the effect of temperature on sodium and potassium permeability (Frankenhaeuser and Moore, 1963) gave the following results. The values of Q_{10} were: for α_m 1.8, for β_m 2.7, for α_h 2.8, for β_h 2.9, for α_n 3.2, for β_n 2.8, for $\overline{P}_{Na}$ 1.3, and for $\overline{P}_K$ 1.2.

The potential V_h (i.e., when $h_\infty = 0.5$) in the node of Ranvier varies from +3 to +20 mV (mean +9.15 mV). At the resting potential, h_∞ varies from 0.59 to 0.95 (mean 0.74).

The complete system of equations describing the behavior of the excitable membrane of the node of Ranvier was suggested by Frankenhaeuser and Huxley (1964):

$$I_m = C \frac{dV}{dt} + I_{Na} + I_K + I_l + I_p \tag{40}$$

$$I_{Na} = P_{Na} \frac{EF^2}{RT} \frac{[Na]_o - [Na]_i \exp \{EF/RT\}}{1 - \exp \{EF/RT\}} \tag{33a}$$

I_k and I_p (the "nonspecific" ionic current; see above) are calculated by means of similar equations for the appropriate concentrations and permeabilities:

$$P_{Na} = \overline{P}_{Na} m^2 h, \tag{34}$$

$$\frac{dm}{dt} = \alpha_m (1 - m) - \beta_m \cdot m \tag{22}$$

and

$$\frac{dh}{dt} = \alpha_h (1 - h) - \beta_h \cdot h, \tag{23}$$

where

$$P_p = \overline{P}_p \cdot p^2, \tag{38}$$

$$\frac{dp}{dt} = \alpha_p (1 - p) - \beta_p \cdot p \tag{39}$$

$$P_K = \bar{P}_K n^2 \tag{36a}$$

$$\frac{dn}{dt} = \alpha_n (1 - n) - \beta_n \cdot n. \tag{19}$$

The values of the rate constants (α and β) for different values of the potential V = E − E_r are given by the following empirically chosen equations:

$$\alpha_h = A(B - V)/\{1 - \exp[(V - B)/C]\} \tag{41}$$

$$\beta_h = A/\{1 + \exp[(B - V)/C]\} \tag{42}$$

$$\alpha_m = A(V - B)/\{1 - \exp[(B - V)/C]\} \tag{43}$$

$$\beta_m = A(B - V)/\{1 - \exp[(V - B)/C]\} \tag{44}$$

$$\alpha_p = A(V - B)/\{1 - \exp[(B - V)/G]\} \tag{45}$$

$$\beta_p = A(B - V)/\{1 - \exp[(V - B)/C]\} \tag{46}$$

$$\alpha_n = A(V - B)/\{1 - \exp[(B - V)/C]\} \tag{47}$$

$$\beta_n = A(B - V)/\{1 - \exp[(V - B)/C]\} \tag{48}$$

where A, B, and C are constants whose standard values are given in Table 1.

TABLE 1. Values of Constants in Equations (41) to (48), 20°C

Rate constant	A ($msec^{-1}$)	B (mV)	C (mV)	Literature citation
α_h	0.1	− 10	6	Frankenhaeuser (1960)
β_h	4.5	+ 45	10	Frankenhaeuser (1963b)
α_m	0.36	+ 22	3 ⎫	Frankenhaeuser (1960)
β_m	0.4	+ 13	20 ⎭	
α_p	0.006	+ 40	10 ⎫	
β_p	0.09	− 25	20 ⎪	Frankenhaeuser (1963a)
α_n	0.02	+ 35	10 ⎪	
β_n	0.05	+ 10	10 ⎭	

Note. $[Na]_0 = 114$ mM; $[K]_0 = 2.5$ mM; $[Na]_i = 13.7$ mM; $[K]_i = 120$ mM; $E_r = -70$ mV; $g_l = 30.3$ mmho/cm^2; $V_l = 0.026$ mV; $C_m = 2$ $\mu F/cm^2$; $P_{Na} = 8 \times 10^{-3}$ cm/sec; $\bar{P}_p = 0.54 \times 10^{-3}$ cm/sec; $\bar{P}_K = 1.2 \times 10^{-3}$ cm/sec. Standard initial data: V = 0 mV; $h_\infty = 0.8249$; $m_\infty = 0.0005$; $p_\infty = 0.0049$; $n_\infty = 0.0268$.

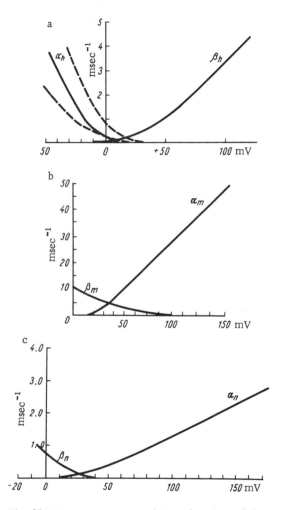

Fig. 28. Rate constants α and β as functions of the membrane potential in the node of Ranvier of Xenopus laevis (a, b, c). Abscissa, change in membrane potential V, in mV; ordinate, rate constants. Broken lines (in graph a) show variations in α_h for different fibers (Frankenhaeuser, 1960; Frankenhaeuser and Huxley, 1964).

The relationships between the rate constants (α and β) and the potential V described by Eqs. (41)-(48) is shown graphically in Fig. 28. The position of the curves relative to the voltage axis is determined by the constants B.

For instance, an increase in the value of B in Eqs. (41)-(48) shifts the α_m versus V, α_n versus V, and α_h versus V curves to the right along the V axis, while a decrease in B shifts these curves in the opposite direction.

The values of the rate constants for given values of B and V are determined by the constants A and C: an increase in A or decrease in C leads to increase in α and β, while a decrease in A or increase in C lowers these rate constants.

The effect of the constants B and A on the parameters of potential and indices of excitability will be examined in Chapters V and VI.

Frankenhaeuser and Huxley (1964) used the above-mentioned system of equations to calculate the behavior of the membrane under normal conditions, when the ionic currents change the membrane potential.

The membrane action potential, calculated in this way, was found to be very similar in its parameters to the experimental spike. Only slight differences were found, but not more than would be expected considering the scatter of the experimental data and inaccuracies of the voltage-clamp method. It was therefore concluded that investigations by the voltage-clamp method can provide a sufficiently complete mathematical description of the currents arising in the medullated fiber, and that there is no need to look for new currents of substantial magnitude in order to understand the mechanism of genesis of the action potential in the node of Ranvier.

The Use of the Mathematical Model to Analyze

the Kinetics of Ionic Currents in Different Phases of

Development of the Action Potential

Modern methods of experimental investigation cannot yet be used to study the kinetics of ionic currents except when the voltage is clamped, i.e., under the usual conditions of action potential generation. All that can be done is to study the changes in the net ionic current (I_i) in relation to the first derivative of the membrane potential with time $\dot{V}$.

We know already that the total current (I_m) flowing through a uniform area of membrane is equal to the sum of the capacitance current $C\dot{V}$, associated with changes in the ion density on the inside and outside of the membrane, and the net ionic current I_i, resulting from diffusion of ions through the membrane:

$$I_m = C\frac{dV}{dt} + I_i. \tag{14}$$

If the stimulus is of very short duration (less than the latent period of the action potential), afters its end I_m in the external circuit becomes equal to zero, and the whole of the ionic current serves simply to charge or discharge the membrane capacitance:

$$-C\frac{dV}{dt} = I_i. \tag{14a}$$

Assuming that the capacitance of the membrane C_m remains constant during action potential generation (Cole and Curtis, 1938), the changes in I_i can easily be calculated from the changes in $\dot{V}$ recorded directly in the experiment.

An example of such a calculation is illustrated in Fig. 29, taken from Dodge (1963). During the period of action of a short stimulus (I_m), the net ionic current I_i is outward in direction, but later it begins to decrease in magnitude, passes through zero at the moment when the local response changes into the ascending phase of the spike, and for a short time becomes negative. The peak of the inward I_i corresponds to the time of the steepest rise of the action potential. The rate of depolarization then decreases, and at the summit of the spike the net ionic current again becomes zero. In the descending phase of the action potential I_i is positive, i.e., it is outward in direction. At the end of the spike it increases slightly, so that repolarization of the membrane is accelerated, after which I_i falls again to zero.

The mathematical model completely reproduces this picture, but at the same time it enables I_i to be subdivided into its com-

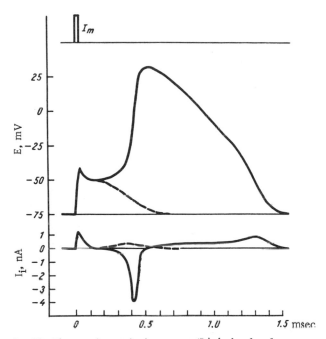

Fig. 29. Changes in net ionic current (I_i) during local reponse
and action potential in a single node of Ranvier during the
action of a short stimulus (I_m) (Dodge, 1963).

poncnts, and it thus provides an explanation for the mechanism of
the changes in V and I observed experimentally.*

Changes in ionic conductances (a) and ionic currents (b), calculated by the equations of Frankenhaeuser and Huxley (1964)
during action potential generation in a node of Ranvier in response
to a very short current pulse, are shown graphically in Fig. 30.
The first feature which will be noticed is that the first zero of I_i
corresponds to the time when the inward sodium current becomes
equal to the outward ionic current $I_l + I_K$. This is the moment of

*Rojas, Bezanilla, and Taylor (1970) have recently developed a method of clamping
the membrane potential at different moments of development of the spike in an internally perfused squid giant axon. In this way they were able to reconstruct the
time course of changes in ionic conductances during generation of the membrane
action potential. Their results showed good agreement with those of calculations
using the Hodgkin-Huxley model (Atwater, Bezanilla, and Rojas, 1970).

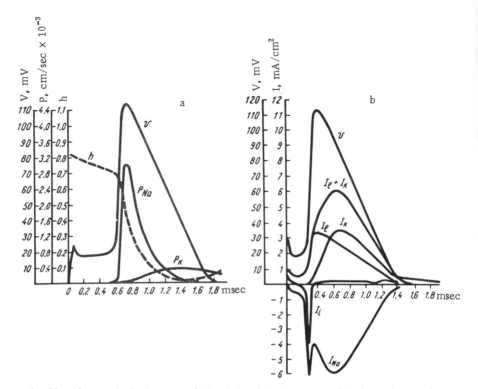

Fig. 30. Changes in ionic permeability (a) and ionic currents (b) of membrane of a Ranvier node during action potential development. Calculated for standard data. Stimulation of short duration (0.05 msec).

critical depolarization of the membrane when the local response begins to grow into an action potential (the mechanism of this process will be examined in more detail in Chapter VI, pp. 153–154).

The steep rise in P_{Na} and I_{Na} clearly coincides with the beginning of the ascending phase of the spike. However, the subsequent course of the changes in P_{Na} and I_{Na} differs substantially. The curve of the changes in P_{Na} has a single maximum, and the maximum of P_{Na} coincides in time with the maximum of the action potential.

By contrast, the curve of the changes in I_{Na} has two maxima, one occurring approximately halfway up the ascending limb of the spike, and the second during the first third of the repolarization phase. The peak of the action potential corresponds to the minimum of the I_{Na} curve in the region of the "saddle" between its two maxima.

What is responsible for these differences in the dynamics of the changes in P_{Na} and I_{Na} ?

The increase in P_{Na} in the ascending phase of the action potential is the direct result of depolarization of the membrane, so that the higher the value of V, the more marked this increase will be. By contrast, I_{Na} depends both on P_{Na} and on the electrochemical potential $V - V_{Na}$, the value of which (it is 120 mV at rest) falls with the development of depolarization (with an increase in V). However, while V is small, the decrease in electrochemical potential is compensated by the simultaneous increase in P_{Na}, and I_{Na} accordingly continues to rise. However, as soon as V comes significantly closer to V_{Na}, the force moving the sodium ions decreases, and the increase in I_{Na} comes to an end. At this moment, the inward net ionic current (I_i), and with it the value of $\dot{V}$, reach their maximum.

In a system containing RC-elements, the changes in potential always lag behind the corresponding changes in the strength of the current. The action potential thus continues to increase for a short time despite the fact that I_{Na}, and with it I_i, have begun to decline.

The increase in V ends only when the density of the outward ionic current ($I_l + I_K$) becomes equal in absolute magnitude to the inward I_{Na}, i.e., when I_i begins to pass through zero. At this moment (corresponding to $\dot{V} = 0$) the spike reaches its maximum.

The end of the increase in V determines the whole subsequent course of development of the action potential. From this moment the outward ionic current ($I_l + I_K$) begins to exceed the inward current (I_{Na}), so that the net ionic current and the changes in membrane potential change their direction, and the process of repolarization begins.

However, the development of this process is somewhat retarded by a second, temporary increase in I_{Na} due to an increase in the electrochemical potential (resulting from the fall in V), while P_{Na} still remains quite high (compare the values of P_{Na} during the first and second peaks of I_{Na}).

It will be clear from Fig. 30a that in the repolarization phase P_{Na} initially falls steeply, and then more gradually. These changes in P_{Na} are due primarily to an increase in the strength of inactivation (a decrease in h) of the depolarized membrane.

Inactivation of the sodium channels in the descending phase of the action potential is not the only cause of the decrease in P_{Na} taking place in this period.

The experiments of Hodgkin and Huxley by the voltage-clamp method showed that there are two different ways in which P_{Na} can be lowered: inactivation of P_{Na} through prolonged depolarization of the membrane and restoration of the original low P_{Na} during a rapid return of the membrane potential to its original resting level (Fig. 24a,b).

These two processes differ in that when P_{Na} is reduced through inactivation, an additional depolarizing stimulus cannot evoke a new significant increase in P_{Na}, whereas in the case of restoration of the initial value of P_{Na}, a new stimulus evokes its activation as usual.

Both these processes take part in repolarization of the membrane in the descending phase of the spike: the increase in P_K and inactivation of P_{Na} lead to a decrease in V, and this in turn lowers the permeability of those sodium channels which were not inactivated, and so on. Hence it follows that in the descending phase of the action potential, a process of regenerative repolarization, like the process of regenerative depolarization in the phase of growth of the spike, develops in the membrane in the descending phase of the action potential.

The role of the nonspecific inward I_p during generation of the action potential in a Ranvier node is very small, for the increase in P_p is small and occurs toward the end of the spike. I_p evidently delays repolarization of the membrane slightly.

The kinetics of the ionic currents in the squid giant axon (Fig. 50) is similar in principle to that occurring in the node of Ranvier. The principal quantitative differences are concerned with the ratios between I_K and I_l (in the giant axon, as in most other excitable structures, excluding nodes of Ranvier, g_l is very small, so that I_l plays a negligible role in spike formation) and in the peak values of I_{Na} and I_K. These last values are almost one order of magnitude below the densities of the corresponding currents in medullated fibers, so that the maximum steepness of the ascending phase of the spike in the giant axon, determined by the ratio between the peak values of I_i and C_m [Eq. (14a)], is only about one-third that in the node of Ranvier (see pages 136-138).

Chapter V

Dependence of Action Potential Parameters on the Constants of Membrane Ionic Permeability

It follows from the examination of the kinetics of the ionic currents accompanying the nervous impulse that all factors inhibiting the increase in sodium permeability must increase the critical level of depolarization and reduce the rate of rise and the amplitude of the action potential. Slowing of sodium inactivation or inhibition of potassium activation, on the other hand, must be accompanied by prolongation of the descending phase of the spike without any appreciable changes in its other parameters.

Experiments in which pharmacological agents have been used or the ionic composition of the fluid bathing the fiber has been changed confirm these theoretical predictions. However, quantitative analysis of the relationship between parameters of the action potential and constants of ionic permeability of the excitable membrane has so far proved difficult on the experimental plane, because most of the agents at present available to the investigator have a complex and multiple action on ionic permeability of the membrane, i.e., they modify several characteristics of the system under investigation simultaneously. The only exceptions are tetrodotoxin and tetraethylammonium, which specifically block sodium and potassium ionic currents.

The investigator who uses mathematical models of an excitable membrane has incomparably wider opportunities, for these models allow any one constant of ionic permeability to be selectively changed, leaving all the other parameters unaltered.

In this chapter I shall describe the results obtained mainly with a model of the membrane of the Ranvier node, because this model has been used most fully to study the problems now being discussed.*

However, before turning to look at this material, I must emphasize the principal difficulty in the way of their interpretation. This difficulty is associated with the fact that the physical meaning of many of the variables and constants which Hodgkin and Huxley suggested has not yet been deciphered, and the possibility accordingly is not ruled out that some of them reflect not so much the character of the molecular mechanism of these processes as the method chosen by these authors for their mathematical description.

For the subsequent analysis of the material we must therefore chiefly use indices of the state of the ionic permeability system of the membrane which can be directly measured under experimental conditions and which are independent of the mathematical methods used to approximate the curves.

These parameters include: the maximal sodium $(\overline{P}_{Na})$ and potassium $(\overline{P}_K)$ permeability, the original level of inactivation $(1 - h_\infty)$, the sensitivity of P_{Na} and P_K to changes in potential, the time constants of sodium activation and inactivation (τ_m, τ_h), and certain others.

On the other hand, the rate constants α_m, β_m, α_h, β_h, α_n, β_n, α_p, and β_p, as well as the constants A, B, and C in Eqs. (41)-(48), may be considered at the present time only as the instruments by means of which changes in the above-mentioned parameters of the membrane actually observed can be simulated under the conditions of this particular model.

Maximal Sodium Permeability $\overline{P}_{Na}$

We examined above how the peak sodium permeability of the membrane varies with the potential (Fig. 21b). The curve of P_{Na} versus V which characterizes this relationship has an S-shaped

---

*The calculations were carried out on a digital computer in collaboration with M. L. Bykhovskii, A. M. Rusanov, R. I. Grilikhes, and A. D. Korotkov in the Laboratory of Cybernetics of the A. V. Vishnevskii Institute of Surgery, Academy of Medical Sciences of the USSR.

rise and a plateau, the height of which corresponds to $\overline{P}_{Na}$ provided that initial inactivation of the sodium system is totally abolished by preliminary strong hyperpolarization of the membrane ($h_{\infty} = 1$).

Investigations on nodes of Ranvier of <u>Xenopus</u> <u>laevis</u> have shown that $\overline{P}_{Na}$ varies in different fibers within wide limits – from 4×10^{-3} to 12×10^{-3} cm/sec. The mean and the standard deviation are $(8.3 \pm 3.6) \times 10^{-3}$ cm/sec.

A number of factors diminishing $\overline{P}_{Na}$ are known: they include tetrodotoxin and general and local anesthetics in high concentrations. A distinguishing feature of tetrodotoxin is that it specifically lowers $\overline{P}_{Na}$ of the membrane while leaving the other constants of its permeability practically unchanged (see page 73).

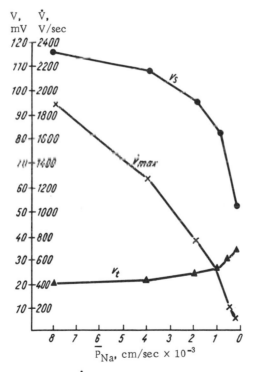

Fig. 31. V_t, $\dot{V}_{max}$, and V_s as functions of initial value of $\overline{P}_{Na}$. All other data standard. Abscissa, $\overline{P}_{Na}$ in 10^{-3} cm/sec. Investigations on model of node of Ranvier (Khodorov, Bykhovskii, Rusanov, and Korotkov, 1966).

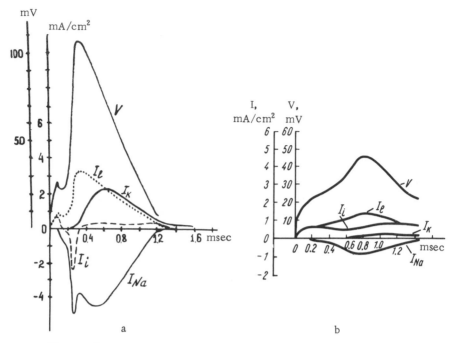

Fig. 32. Membrane action potentials and ionic currents in single node of Ranvier calculated for a decrease in $\overline{P}_{Na}$: a) $\overline{P}_{Na}$ reduced by 50% (to 4×10^{-3} cm/sec). Short stimulus, above threshold; b) $\overline{P}_{Na}$ reduced by 27 times. Stimulation by steady current of 0.7 mA/cm².

There are grounds for considering that tetrodotoxin directly blocks the sodium channels in the membrane (see page 267) and that $\overline{P}_{Na}$ is thus proportional to the total number of active (i.e., sensitive to changes in membrane potential) sodium channels. *
It was therefore interesting to analyze the way in which the various parameters of the action potential depend on the inital value of $\overline{P}_{Na}$.

The relationships between $\overline{P}_{Na}$ and the values of the threshold potential V_t, the maximal rate of rise $\dot{V}_{max}$, and amplitude of the action potential V_s are shown graphically in Fig. 31. Changes in shape of the spikes calculated under these conditions and of their accompanying ionic currents are shown in Fig. 32.

---

*It is assumed that each channel can exist in only two states: open and closed, and that different channels have different thresholds.

The first fact to be noticed is that a decrease in the maximum sodium permeability to 50% of its initial value (i.e., from 8×10^{-3} to 4×10^{-3} cm/sec) has very little effect on the threshold potential* and the spike amplitude, but it reduces its maximal rate of rise ($\dot{V}_{max}$) considerably.

The reason for this sensitivity of $\dot{V}_{max}$ to changes in $\overline{P}_{Na}$ is because the value of $\dot{V}_{max}$ is directly proportional to the strength of the inward sodium current which, in turn, is determined by the degree of increase in sodium permeability during depolarization.

A substantial fall in amplitude of the action potential and a corresponding rise in the threshold potential are found only when $\overline{P}_{Na}$ is reduced to 4-6% of its initial value. Hence, at $\overline{P}_{Na} = 0.3 \times 10^{-3}$ cm/sec, the calculated response acquires an intermediate form between a spike and a local response (Fig. 32b).

Steady-State Sodium Inactivation

$(1 - h_\infty)$ and Time Constant τ_h

Assuming that $\overline{P}_{Na}$ is actually proportional to the number of active sodium channels in the area of membrane investigated, the initial level of inactivation ($1 - h_\infty$; see page 84) can be regarded as a measure of the proportion of these channels inactivated at a given resting potential. In the same way, h_∞ is a measure of the proportion of the channels accessible to activation during the action of a depolarizing stimulus.

The standard value of h_∞ at a resting potential of -70 mV, accepted for the model of the membrane of the node, is 0.8249 (Frankenhaeuser and Huxley, 1964). This value can be increased or decreased by changing the rate constants α_h and β_h, since

$$h_\infty = \frac{\alpha_h}{\alpha_h + \beta_h}$$

Changes in α_h and β_h (or more precisely, a change in their dependence on the membrane potential), on the other hand, can

---

*In all these investigations, the value of V measured when $\dot{V} = 80$ V/sec was chosen conventionally (see page 163) as the index of the threshold potential V_t. Since the initial level of the membrane potential of the model in all calculations was constant (-70 mV), the changes in threshold potential (v_t) accurately reflect changes in the critical potential E_c.

be obtained on a mathematical model by a corresponding change in the constant B in Eqs. (41) and (42) (Khodorov, Bykhovskii, Rusanov, and Korotkov, 1966).

It is stated in the literature that the values of α_h and β_h can vary independently in different nerve fibers of amphibians, and that the variations in the rate constant α_h are particularly large (Frankenhaeuser, 1960; Vallbo, 1964) (Fig. 28a). Accordingly, in the present writer's investigations with a mathematical model, the initial level of inactivation was changed by varying each of these constants in isolation.

Various parameters of the action potential, calculated as a function of the initial level of h_∞, changes in which were produced

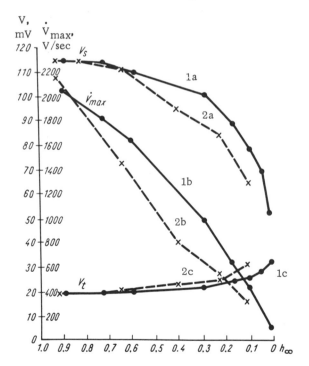

Fig. 33. Amplitude of action potential (V_s), maximal rate of its rise ($\dot{V}_{max}$), and threshold potential (V_t) as functions of steady-state inactivation (h_∞). Curves 1a, 1b, and 1c were obtained by changing h_∞ through the rate constant α_h; curves 2a, 2b, and 2c were obtained by changing the rate constant β_h (Khodorov, Bykhovskii, Rusanov, and Korotkov, 1966).

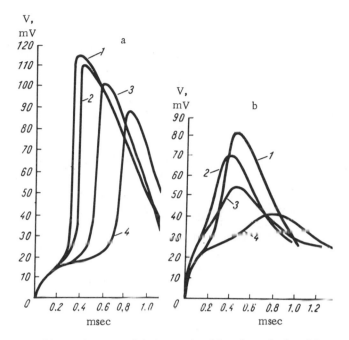

Fig. 34. Action potentials in a node of Ranvier calculated for
different initial levels of inactivation. Changes in h_∞ produced
through changes in the constant B in Eq. (41) describing the dependence of α_h on V. a) Curves 1, 2, 3, and 4 plotted for the
following values of h_∞: 0.886, 0.599, 0.170, and 0.172. Stim
ulating current 0.5 mA/cm²; b) responses 1, 2, and 3 at values
of h_∞ of 0.093, 0.048, and 0.023 respectively. Stimulating
current 1 mA/cm². Curve 4 is the same as curve 3 but for a
weaker current, namely 0.75 mA/cm² (Khodorov, Bykhovskii,
Rusanov, and Korotkov, 1966).

through the constants α_h (curves 1a, 1b, 1c) and β_h (curves 2a,
2b, 2c), are shown graphically in Fig. 33.

Clearly an increase in the initial inactivation of the membrane
(a decrease in h_∞) leads to an increase in the threshold potential
V_t and a decrease in $\dot{V}_{max}$ and V_s. The total duration of the spike
is also reduced.

With a very considerable decrease in h_∞, the action potentials
come to resemble local responses (Fig. 34). A similar change
takes place in the spike if h_∞ is reduced by a change in the rate
constant β_h.

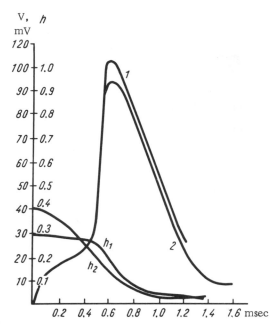

Fig. 35. Action potentials of a node calculated for an increase in the initial sodium inactivation through changes in the rate constants α_h and β_h. 1) h_∞ reduced to 0.29 through a change in the constant α_h; 2) h_∞ reduced to 0.4 through a change in the constant β_h. Stimulating current in both cases 0.5 mA/cm². Curves h_1 and h_2 show kinetics of increase in inactivation during development of action potential in cases 1 and 2 (Khodorov, Bykhovskii, Rusanov, and Korotkov, 1966).

However, with an equal decrease in h_∞, i.e., with the same level of inactivation, changes in all the parameters of the action potential are more sharply defined if the change in h_∞ takes place through changes in the constant β_h. This is explained by different effects of the constants α_h and β_h on the rate of development of the inactivation process. The time constant of this process $\tau_h = 1/(\alpha_h + \beta_h)$, so that with a decrease in α_h (at a given value of V), τ_h is increased, i.e., the development of inactivation is retarded, but with an increase in β_h, the time constant τ_h is reduced and, consequently, the inactivation is accelerated.

The meaning of these differences in the effects of changes in α_h and β_h becomes particularly clear if the results shown graphically in Fig. 35 are examined. Two action potentials and the corresponding curves showing how h changes with time are also shown in Fig. 35. The first spike was obtained when the initial value of h_∞ was reduced to 0.41, i.e., by 50% of the standard value, through a change in the constant β_h, and the second spike with a decrease in h_∞ to 0.29 (i.e., to 35% of the initial value) by a change in the rate constant α_h. Despite the greater decrease in h_∞, in the second case V_s and V_{max} of this action potential are higher than for the first spike. The reason is that with an increase in β_h the time constant τ_h is shortened and the rapidly developing inactivation appears to cut short the growth of the action potential. With a very considerable increase in β_h, the calculated action potentials of the node become graded. It thus becomes clear that the amplitude and rate of rise of the action potential depend not only on the initial level of inactivation $(1 - h_\infty)$, but also on the rate of development of this process during membrane depolarization, i.e., on the value of τ_h.

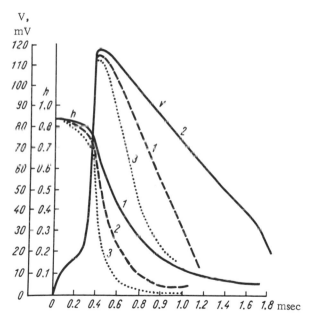

Fig. 36. Effect of change in time constant of sodium inactivation (τ_h) on action potential of node of Ranvier. 1) Action potential and changes in variable h during generation of action potential calculated for standard data for the model; 2 and 3) action potential and changes in h with a twofold increase and decrease in the time constant τ_h. These changes were produced by a simultaneous increase or decrease in the values of the constants A in Eqs. (41) and (42). $I_m = 0.5$ mA/cm^2.

However, τ_h has the greatest effect on the duration of the descending phase of the action potential.

This is clearly seen in the case of isolated changes in this parameter with a change in the time constant. Such a change can be obtained in the model by simultaneously increasing or decreasing the constants A in Eqs. (41) and (42).

Calculated changes in action potential with a twofold change in the value of τ_h are shown graphically in Fig. 36. It will be clearly seen how slowing the inactivation process leads to a small increase in the amplitude of the spike and to a very considerable increase in the duration of its descending phase. A decrease in τ_h, on the other hand, while h_∞ remains at the same initial level, gives rise to the opposite effect: a slight decrease in amplitude of the action potential and considerable shortening of the repolarization phase (curve 3).

In experiments to test the action of pharmacological agents or ions on the nerve fiber, the investigator is frequently faced with the problem of how to differentiate between depression of an action potential due to inactivation and the similar change due to a decrease in maximum sodium permeability of the membrane $\overline{P}_{Na}$.

The result of the calculations shown in Fig. 37 provides a solution to this problem. In the case of a decrease in h_∞ (Fig. 37a), the action potential can be completely restored by hyperpolarization of the membrane by 30 mV, i.e., by a decrease in V to -30 mV.

Changes in the action potential caused by reducing $\overline{P}_{Na}$ are not, however, reversed by hyperpolarization (Fig. 37b). The slight increase in the response arising during hyperpolarization in this case is connected with a decrease in the resting inactivation which took place in the initial state of the model.

All these results shed considerable light on the normal variations in threshold, amplitude, and shape of the potentials in different nerve fibers, and also on the analysis of their changes under the influence of pharmacological agents which do not depolarize the excitable membrane. These agents include local anesthetics (procaine, cocaine, xylocaine, etc.), general anesthetics (urethane, nembutal, ether, etc.), and also poisons (tetrodotoxin, saxitoxin, etc.).

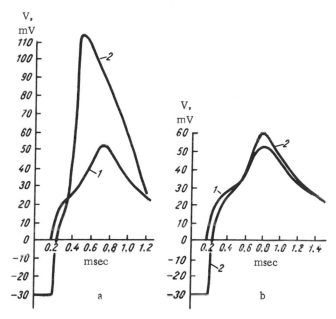

Fig. 37. Difference between effects of membrane hyperpolarization when action potential is depressed by sodium inactivation (a)
and when the maximum sodium permeability $\overline{P}_{Na}$ is lowered (b).
a) Curve 1 calculated for a decrease in the constant D in Eq.
(42) from 45 to 5 mV (h = 0.12); curve 2 shows the same after
removal of the inactivation by hyperpolarization of the membrane by V = −30 mV; b) curve 1 calculated for a decrease in
$\overline{P}_{Na}$ from 8 × 10^{-3} to 0.3 × 10^{-3} cm/sec; 2) the same, after
hyperpolarization by −30 mV (Khodorov, Bykhovskii, Rusanov,
and Korotkov, 1966).

Procaine (10^{-4} g/ml) and tetrodotoxin (10^{-8} g/ml), if applied
to single nodes of Ranvier from frog's nerve fibers in concentrations sufficient to block the action potential, sharply inhibit the increase in sodium permeability of the membrane during depolarization, but leave the resting potential of the fiber virtually unchanged
(Dettbarn et al., 1960; Khodorov and Belyaev, 1965b, 1967; Hille,
1966, 1967).

These agents have the same action on the membrane of cuttlefish giant axons (Taylor, 1959; Nakamura et al., 1965; Takata et
al., 1966) and skeletal muscle fibers (Narahashi et al., 1964; Khodorov and Vornovitskii, 1967).

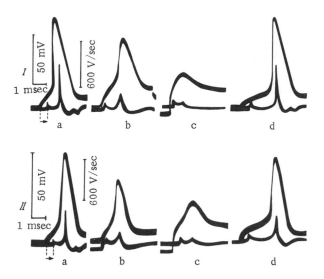

Fig. 38. Effect of tetrodotoxin (TTX) (I) and of procaine (II)
on electrical activity of node of Ranvier from isolated frog
(R. temporaria) nerve fiber. Top line shows change in potential V; bottom line shows $\dot{V}$ in V/sec. For convenience of
reading the records, the beginning of the changes in $\dot{V}$ in all
figures is shifted slightly to the right relative to the beginning
of the changes in V (the extent of the shift is shown by the
broken lines and the arrow). I: a) node in Ringer's solution of
normal composition; b) 10^{-9} g/ml TTX; c) 10^{-8} g/ml TTX;
d) recovery of action potential after rinsing out TTX. II:
the same as I, but after the addition of procaine in different
concentrations; a) initial action potential; b) 2×10^{-5} g/ml
procaine; c) 10^{-4} g/ml procaine; d) incomplete recovery
after rinsing node in Ringer's solution (Khodorov and Belyaev,
1967).

Changes in the amplitude and shape of the potential caused by
procaine and tetrodotoxin are outwardly very similar (Fig. 38a,b).
However, as investigations showed, hyperpolarization of the membrane restores generation of action potentials only in the case of
treatment with procaine (Fig. 39A). The effects of tetrodotoxin
are not reversed by hyperpolarization (Fig. 39B).

Since $\dot{V}_{max}$ in the fiber treated with procaine, when exposed
to strong and prolonged (of the order of 1 sec) hyperpolarization
of the membrane, reaches values close to the initial level obtained
during similar hyperpolarization (compare d' and b'), it was con-

cluded that the effects of procaine in these doses are attributable
mainly to inactivation of the sodium permeability of the membrane.

By contrast, the inhibitory action of tetrodotoxin is due to a
decrease in the maximum sodium permeability $\overline{P}_{Na}$ (Khodorov and
Belyaev, 1967; Khodorov and Vornovitskii, 1967).

Hyperpolarization of the membrane by a steady current is not
the only factor which removes the sodium inactivation of the membrane.

Ca^{++} ions have the same effect. An increase in their concentration in the solution increases the resistance of the nerve membrane to the inactivating action of prolonged subthreshold depolar-

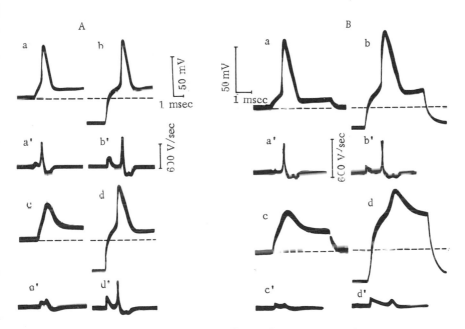

Fig. 39. Action of hyperpolarization on effects of procaine (A) and tetrodotoxin (B).
a, b, c, d) Changes in membrane potential (V) of Ranvier node; a', b', c', d') corresponding changes in $\dot{V}$. Broken line shows initial level of resting potential. A:
a, a' and b, b') control responses of node in Ringer's solution before and during action
of hyperpolarizing current (respectively); c, c', d, d') node in 10^{-4} g/ml procaine
solution. Hyperpolarization of the membrane restores action potential generation.
B: a, a' and b, b') control response of node in Ringer's solution; c, c' and d, d') node
in 10^{-8} g/ml tetrodotoxin solution. Hyperpolarization does not restore excitability.
Values of membrane potential not corrected (Khodorov and Belyaev, 1967, with modifications).

ization (Frankenhaeuser, 1957b; Khodorov, 1964, 1965; Khodorov and Belyaev, 1966) and of local anesthetics (Khodorov and Belyaev, 1965; Blaustein and Goldman, 1966a,b).

Calcium ions also have a definite restorative effect when they act on other excitable structures altered by procaine.

Aceves and Machne (1963) observed this effect on neurons of spinal ganglia, and Vornovitskii and Khodorov (1966) observed it on frog skeletal muscle fibers (Fig. 40).

An even stronger restorative action on the inactivated membrane of the medullated nerve fiber is shown by Ni^{++}, Co^{++}, and Cd^{++} ions.

Addition of these ions to Ringer's solution in concentrations of the order of 0.5-1 mM restores action potentials suppressed by procaine, cocaine, urethane, an excess of K^+ ions in the solution, a cathodic current, or mechanical alteration of the node of Ranvier

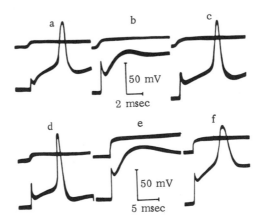

Fig. 40. Effect of changes in Ca^{++} ion concentration in solution on effects of procaine: a) action potential of a skeletal muscle fiber in medium with normal Ca^{++} ion concentration; b) during action of 5×10^{-4} g/ml procaine; c) combined action of 5×10^{-4} g/ml procaine and 27 mM $CaCl_2$; d) action potential in normal salt medium; e) muscle in solution of 0.3×10^{-4} g/ml procaine and with lowered Ca^{++} ion concentration (to 0.18 mM); f) muscle in solution of 0.18 mM $CaCl_2$ without procaine (Vornovitskii, 1966).

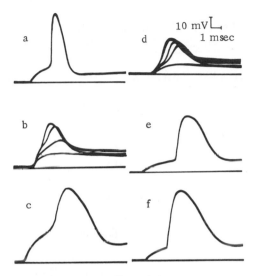

Fig. 41. Restorative effect of nickel ions on action potential generation in a single node of Ranvier from an isolated frog nerve fiber blocked by
procaine (a-c) and injured mechanically (d-f).
a) Initial action potential of node in Ringer's ·
solution of normal composition; b) after treatment with 10^{-4} g/ml procaine; c) after addition of 1 mM $NiCl_2$ to procaine solution; d) node
injured during dissection; e) 60 sec after addition of 1 mM $NiCl_2$ to Ringer's solution; f) the
same, after 300 sec (Khodorov and Belyaev,
1964a).

(Takahashi, Usuda, and Ehara, 1962; Khodorov and Belyaev, 1963b,
1964a,b) (Fig. 41) (see also page 290).

However, these ions possess the following important property
which distinguishes them from Ca^{++} ions: they cause a marked
increase in the duration of the action potential (Tasaki, 1959a;
Spyropoulos and Brady, 1959; Khodorov and Belyaev, 1963b, 1964a,b;
Meves, 1963; Meves and Weymann, 1963).

This effect is observed not only on intact nerve fibers, but
also when Ni^{++} ions act on altered nodes of Ranvier, in which case
the restored spikes are considerably prolonged (Fig. 41c,e,f). The
reason for this phenomenon is that Ni, Co, and Cd not only weaken
the original inactivation of the membrane (increase h_{∞}), but they
also considerably prolong the time constants τ_m, τ_n, and τ_h (see

page 138). The theoretical effects of an increase in each of these time constants are shown in Figs. 36, 44, and 49b).

Reactivity of the Sodium System

(Sensitivity of P_{Na} to Changes in

Membrane Potential)

Investigations by the voltage-clamp method have shown that when certain influences are brought to bear on the nerve fiber and, in particular, during changes in the concentration of divalent cations in the medium, the P_{Na} versus V curve can be shifted along the voltage axis even if the maximum sodium permeability ($\overline{P}_{Na}$) and the initial level of inactivation ($1 - h_\infty$) are unchanged.

This means that the degree of the increase in sodium permeability associated with a given depolarization depends not only on the total number of sodium channels free from inactivation, but also on the sensitivity of the mechanism "opening" these channels to depolarization.

The shift of the potential (φ_{Na}) at which the peak value of P_{Na} rises to half the maximum value (Fig. 42) can be used as the index of this sensitivity (reactivity).

An increase in φ_{Na}, corresponding to a shift of the P_{Na} versus V curve to the right (curve 2), implies a decrease in reactivity of the sodium system, whereas a decrease in φ_{Na} (curve 1) indicates that this reactivity is increased.

In model experiments these shifts of the P_{Na} versus V curve can be obtained by increasing or decreasing the constants B in Eqs. (43) and (44) describing the rate constants α_m and β_m as a function of membrane potential (see pages 95-96).

The curve of the threshold potential V_t, the maximum steepness of the ascending phase of the spike $\dot{V}_{max}$, its total duration t (measured at the height of 0.3 of its amplitude starting from its peak), and the maximum of P_{Na} at the peak of the action potential,*

---

* The maximum of P_{Na} at the peak of the action potential ($\dot{P}_{Na}$) must not be confused with the maximum sodium permeability $\overline{P}_{Na}$, measured during fixed depolarization of the membrane. The first of these indices is always much smaller than the second, for during the short time of development of the spike, P_{Na} is unable to rise to its limit $\overline{P}_{Na}$.

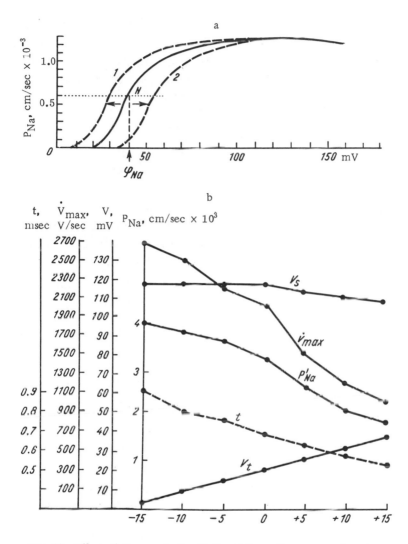

Fig. 42. Effect of changes in sensitivity of the sodium system (φ_{Na}) to depolarization on parameters of the action potential of a model of the node of Ranvier: a) shift of P_{Na} versus V curve (compare Fig. 21b) with an increase (1) and decrease (2) in φ_{Na}; N) P_{Na} versus V curve obtained under voltage-clamp conditions with standard values of constants; b) changes in threshold potential (V_t), amplitude of action potential (V_s), maximum steepness of its ascending phase ($\dot{V}_{max}$), sodium permeability at maximum of spike (P'_{Na}), and its total duration (t) with an increase and decrease in the value of φ_{Na}. Abscissa: changes in φ_{Na} ($\Delta\varphi_{Na}$) from initial value ($\Delta\varphi_{Na} = 0$) (Bykhovskii, Khodorov, Rusanov, and Korotkov, 1967, with modifications).

as functions of the shift of φ_{Na}, either increasing ($\Delta\varphi_{Na} > 0$) or decreasing ($\Delta\varphi_{Na} < 0$), are shown in Fig. 42b.

Clearly an increase in φ_{Na} leads to an increase in the threshold potential, a decrease in the maximum steepness of the spike, and a slight lowering of the level of its maximum. The spike duration is shortened. With a decrease in φ_{Na}, all these parameters of the spike are changed in the opposite direction.

Let us examine these changes in more detail.

The Threshold Potential

On the basis of his results obtained by the voltage-clamp method Frankenhaeuser (1960) postulated that the normal variations of the threshold potential (V_t) of nodes of Ranvier from different nerve fibers are determined principally by the constants α_m and β_m.

Vallbo (1964b) attempted to find a quantitative relationship between the threshold potential of different nerve fibers and the constants α_m and β_m, and in order to do so he combined the method of clamping the membrane potential with the ordinary method of stimulation and recording the potentials from single nodes of Ranvier. As the index of the threshold potential V_t he chose the value of V when $\dot{V} = 80$ V/sec, and as the index of the state of the sodium system he chose the potential (V_a) at which the peak of the initial sodium current reaches half its maximum value. The correlation between V_t and the threshold potential was found to be very close ($r = 0.86$; $P < 0.001$). The coefficient of regression was 0.56.

It was concluded from these facts that the critical potential of intact fibers is dependent mainly on the position of the I_{Na} versus V curve relative to the voltage axis. The higher the value of I_{Na} for a given value of V, the lower the critical potential, and vice versa. Since the position of the I_{Na} versus V curve is determined by the rate constants of activation of sodium permeability, it was natural to conclude that the critical potential is closely connected with the values of these constants.

Calculations using the model completely confirmed this hypothesis (Frankenhaeuser and Vallbo, 1965; Bykhovskii, Khodorov, Korotkov, and Rusanov, 1967).

It will be clear from Fig. 42b that V_t and, consequently, the critical potential E_c (since the resting potential is unchanged) are

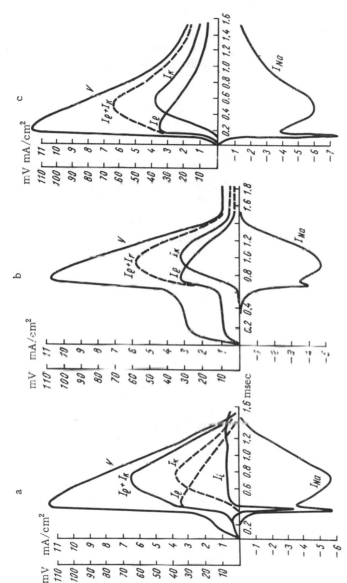

Fig. 43. Calculated changes in ionic currents in a node of Ranvier during the action potential with shifts in the reactivity of the sodium system a) standard values; b) with a decrease and c) with an increase in reactivity of the sodium system ($\Delta \varphi_{Na} = +15$ mV and -15 mV). Changes in φ_{Na} were produced by a simultaneous increase or decrease in the values of the constant B in Eqs. (43) and (44) by 15 mV; a) stimulation with steady current $I_m = 0.5$ mA/cm², b) 0.3 mA/cm²; c) spike in response to stepwise decrease in φ_{Na} by 15 mV.

linear functions of the sensitivity of the sodium system to depolarization, and that each increase of 10 mV in the value of φ_{Na} corresponds to an increase of the same magnitude in the threshold potential.

It can thus be seen that the critical depolarization level of an
excitable membrane is a function of the sensitivity of its sodium
system to depolarization.

Maximal Steepness of Rise of the Ascending
Phase of the Action Potential

A decrease in the sensitivity of the sodium activating system
to depolarization not only increases the threshold potential, but
also produces a definite slowing of the rate of rise of the action
potential (Fig. 42b, $\dot{V}_{max}$).

The direct cause of this effect is weakening of the first I_{Na}
peak (Fig. 43a,b) which, as has already been stated (see page 101
is responsible for recharging the membrane during generation of
the nervous impulse.

With an increase in sensitivity (a decrease in φ_{Na}) the first
peak of the sodium current, on the contrary, is considerably increased (Fig. 43c) compared with its initial value (Fig. 43a), and
the steepness of rise of the action potential thus also increases.

Amplitude of the Action Potential

As Fig. 42b shows, changes in φ_{Na} have little effect on spike
amplitude. By contrast, the maximum reached by P_{Na} at the peak
of the action potential (P'_{Na}) undergoes substantial changes under
the influence of changes in φ_{Na}.

What is the reason for the decrease in the maximum P'_{Na} at
the peak of the action potential with an increase in the value of
φ_{Na}?

The maximal sodium permeability ($\overline{P}_{Na}$) being constant, this
can take place only through a lesser activation of the sodium channels in the membrane (a smaller degree of increase in the variable
of activation m) or through their greater inactivation. Computa-
tion showed that when the action potential reaches its maximum,
the variable m has risen to about the same value (0.84), both when

the reactivity of the sodium system is normal (standard) and when it is lowered by 15 mV ($\Delta\varphi_{Na} = 15$ mV). The value of h, on the other hand, at the maximum of the spike is 0.56 in the first case, but in the second it is reduced to 0.30. This means that the chief cause of the decrease in the maximum of P_{Na} during the action potential is the more pronounced inactivation when φ_{Na} is increased. At first glance this conclusion may appear unexpected, but it will be fully understood if it is remembered that the increase in φ_{Na} raises the threshold, and with a high threshold the value of h is able to fall much more (to 0.53) during the period of increase of V up to the critical level than with the original values of φ_{Na} and V_t (to 0.792).

Duration of the Action Potential

Despite the fact that the steepness of the ascending phase of the action potential falls with an increase in φ_{Na} and rises with a decrease in φ_{Na}, the total duration of the spike in the first case is reduced, and in the second it is increased. This conclusion is valid whether the spike duration is measured at the level of the critical potential or at a somewhat lower level, namely 0.3 of the amplitude of the active part of the spike.

Changes in the duration of the action potential are clearly connected with the steepness of the ascending phase of the spike, i.e., with the rate of development of repolarization. At normal and, in particular, at raised values of φ_{Na} the rate of membrane repolarization increases as V approaches the level of 40-50 mV (the "shoulder" on the descending part of the spike). By contrast, at low values of φ_{Na}, i.e., when the reactivity of the sodium system is high, the rate of repolarization at V ~ 40-50 mV falls distinctly, and the descending phase of the spike becomes more sloping (Fig. 43c). At the same time, recovery of P_{Na} to its original low value is retarded. It was stated above (see page 102) that two factors are responsible for the decrease in P_{Na} during the repolarization phase: inactivation of P_{Na} and "closure" of the sodium channels arising when V approaches the initial level of the resting potential. At low values of φ_{Na} this second process is retarded, because even weak depolarization of the membrane is sufficient to maintain P_{Na} relatively high.

These results are important for the understanding of some aspects of the action of divalent cations and, in particular, of Ca^{++}

ions which, besides weakening inactivation (shifting the h_∞ versus V curve toward higher values of V), also bring about a characteristic decrease in the reactivity of the sodium (Frankenhaeuser and Hodgkin, 1957; Hille, 1968a; Blaustein and Goldman, 1966b).

Since an increase in $[Ca]_0$ weakens inactivation of the sodium system of the membrane, while a decrease in $[Ca]_0$ strengthens its inactivation, it is best to examine the effects of an excess and deficiency of Ca^{++} ions under the conditions of complete abolition of inactivation   (i.e., with an increase in h_∞ to 1.0). This can be done by producing strong hyperpolarization of the membrane by an anodic current.

Experiments show that after preliminary hyperpolarization the effect of Ca^{++} ions on the critical potential becomes particularly conspicuous. The critical potential in nodes of Ranvier is an approximately linear function of the logarithm of Ca^{++} concentration. The steepness of the ascending phase of the action potential rises if there is a deficiency of Ca^{++} ions in the medium, the amplitude increases to its original maximum, and the duration of the spike becomes longer than initially. An excess of Ca^{++} ions against the background of hyperpolarization of the membrane has little effect on the amplitude of the action potential but reduces its steepness slightly. The duration of the spike is shortened (Frankenhaeuser, 1957b; Ulbricht 1964; Khodorov, 1965a,b).

All these changes in the threshold and shape of the action potential under the influence of Ca^{++} ions and under conditions when inactivation of the sodium system is abolished or is slight are very similar in character to those obtained by calculation during the corresponding change in the reactivity of the sodium system to depolarization.

In fact, it was discovered that an increase in φ_{Na}, like an excess of Ca^{++} ions, increases the threshold potential, reduces the steepness of the ascending phase of the action potential, and accelerates repolarization (especially as V approaches its initial level).

A decrease in φ_{Na}, on the other hand, lowers the threshold V_t, increases the steepness of the ascending phase of the spike, and increases the duration of the repolarization phase, i.e., gives rise to changes in the action potential similar to those observed in response to a decrease in $[Ca]_0$.

Analysis of the action of Ca^{++} ions in the case of marked inactivation of the sodium system, i.e., with a low initial value of h_∞, is more complex. In this case, under the influence of Ca^{++} ions not only is φ_{Na} increased, but h_∞ is also increased. However, the increase in the previously lowered value of h_∞ leads to a change in the parameters of the action potential which is directly opposite to that produced by an increase in φ_{Na}. Whereas, in fact, an increase in φ_{Na} leads to an increase in the threshold potential, and to a slight decrease in the maximum steepness and amplitude of the spike, weakening of inactivation creates a tendency for the previously increased V_t to diminish, and for $\dot{V}_{max}$ and the amplitude of the action potential to increase (to recover).

Thus the calculations show that the end result of simultaneous shifts in φ_{Na} and in h_∞ induced by the changes in $[Ca]_0$ depends on the initial level of h_∞.

If h_∞ is moderately reduced (to 0.7-0.6), its increase to 1.0 will not appreciably affect the parameters of the action potential (Fig. 33). In that case changes in the reactivity of the sodium system, i.e., changes in φ_{Na}, assume the leading role. The result obtained in this case simulates well the effect of Ca^{++} ions on generation of the action potential in a nerve fiber under normal physiological conditions.

If, on the other hand, sodium inactivation is high (h_∞ is small), and in the case of a simultaneous increase in h_∞ and φ_{Na}, the change in h_∞ becomes dominant. This leads to a marked increase in the previously low values of the maximum steepness and amplitude of the action potential. The main result of an increase in φ_{Na} is that the restored action potential begins at a higher critical level of depolarization.

This effect in fact also takes place through the action of Ca^{++} ions on nerve fibers when treated with procaine or by some other agent inactivating sodium permeability (see page 118).

These results also provide a satisfactory explanation of the effects of a decrease in Ca^{++} ion concentration in the medium. If sodium inactivation of the membrane of a nerve fiber is slight or is artificially maintained at a low level by a hyperpolarizing current, a decrease in $[Ca]_0$ causes changes in the action potential due mainly to a decrease in φ_{Na}.

Under these circumstances the critical potential E_c falls so sharply that "spontaneous" discharges of impulses arise in nerve or muscle fibers.

The generation of "anode-break excitation" in nerve fibers in a medium with reduced Ca^{++} ion concentration results from a similar mechanism (Frankenhaeuser and Widen, 1956): hyperpolarization of the membrane abolishes sodium inactivation, as a result of which the critical level of depolarization is reduced to the initial level of the resting potential. It is now sufficient simply to break the anodic current and to return the membrane potential to its initial level for an action potential to be generated.

If, however, h_∞ is lowered (for example, as the result of prolonged preliminary depolarization of the membrane, the action of local anesthetics or narcotics, mechanical injury to the membrane, and so on), the dominant effect during a decrease in $[Ca]_0$ is the change in h_∞, and as a result generation of the action potential in the nerve fiber is strongly inhibited (Khodorov and Belyaev, 1965b, 1966b, 1967).

Time Constant of Sodium Activation τ_m

The time constant of sodium activation τ_m determines the rate of the changes in P_{Na} of the membrane during its depolarization and repolarization.

Isolated changes in τ_m, while all other characteristics of the system remain constant, can be obtained in the model by increasing or decreasing the constants A in Eqs. (43) and (44), describing α_m and β_m as functions of the membrane potential.

Calculations showed that with a twofold increase of τ_m, only a very small increase in the threshold potential (about 1 mV) is observed (see page 172), but the latent period of spike generation is considerably prolonged (up to 3 msec for a current of rheobase strength) and the steepness of its rising phase is considerably reduced (Fig. 44).

It is interesting that simultaneously with the decrease in rate of rise of the spike, the rate of the repolarization is also appreciably reduced. The reason for this effect is that with an increase in τ_m, not only is activation of the sodium channels slowed, but so also is their reversion to the resting state as the membrane potential comes closer to its initial value.

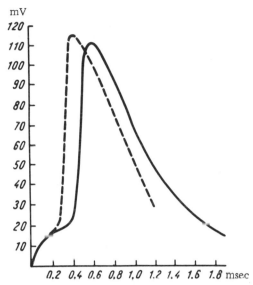

Fig. 44. Effect of an increase in the time constant of sodium activation τ_m on the action potential in a node of Ranvier. Broken line shows action potential calculated for standard values of the
constants, continuous line the same for a twofold
increase in the time constant of sodium activation.
This change in τ_m was produced by doubling the
values of the constants A in Eqs. (43) and (44) sim
ultaneously.

Maximal Potassium Permeability $\overline{P}_K$

The maximal potassium permeability $\overline{P}_K$ can be measured experimentally during strong, fixed depolarization of the membrane.
It is apparently determined by the total number of electrically excitable potassium channels present in the membrane.* According
to Vallbo (1964), the value of $\overline{P}_K$ for nerve fibers of Xenopus laevis
is $(0.93 \pm 0.19) \times 10^{-3}$ cm/sec.

Variation of $\overline{P}_K$ between 0.6×10^{-3} and 2.4×10^{-3} cm/sec has
virtually no effect on the threshold potential of the model. This
calculated result is perfectly easy to understand if it is remembered
that no significant change can take place in $\overline{P}_K$ while the membrane
potential is rising to its critical value and, consequently, it cannot
have any appreciable effect on the magnitude of the critical potential. $\overline{P}_K$ has a significant effect only on the descending phase

---

*See footnote on page 108.

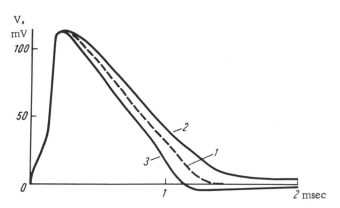

Fig. 45. Effect of an increase and decrease in the maximum potassium permeability on the shape of the action potential. Calculated: 1) for standard value $\overline{P}_K = 1.2 \times 10^{-3}$ cm/sec; 2) for $\overline{P}_K = 0.6 \times 10^{-3}$ cm/sec; 3) for $\overline{P}_K = 2.4 \times 10^{-3}$ cm/sec (Frankenhaeuser and Huxley, 1964).

of the action potential: an increase in $\overline{P}_K$ shortens this phase, while a decrease in $\overline{P}_K$ lengthens it (Fig. 45).

Reactivity of the Potassium System (Sensitivity of $\overline{P}_K$ to Changes in Membrane Potential)

The reactivity of the potassium system can be judged from the change in membrane potential required to increase the potassium permeability to 50% of its maximum value $\overline{P}_K$. This change will subsequently be represented by the symbol φ_K (Fig. 46a), and its decrease or increase relative to its initial value will be represented by $-\Delta\varphi_K$ and $+\Delta\varphi_K$ respectively.

When working with the model, these changes in φ_K can be produced by simultaneously decreasing or increasing the values of the constants B in Eqs. (47) and (48).

Effects of an Increase in Reactivity of the Potassium System

A decrease in φ_K leads to an increase in the outward potassium current during development of the action potential.

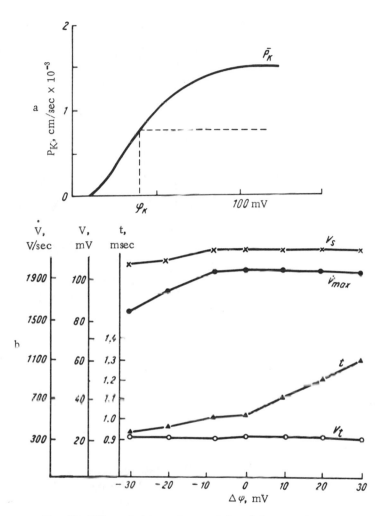

Fig. 46. Effect of changes in reactivity of the potassium system on parameters of the calculated action potential in a single node of Ranvier: a) $\bar{P}_K$ as a function of the fixed change in membrane potential toward depolarization in a node of Ranvier. Abscissa, magnitude of potential change; ordinate, $\bar{P}_K$ measured during the period when I_K reaches a stationary level (Frankenhaeuser, 1963). Initial potential $V = 0$; b) calculated changes in amplitude of action potential V_s, maximum steepness of its ascending phase $\dot{V}_{max}$, duration t, and threshold potential V_t with an increase (on the right) and decrease (on the left) in φ_K.

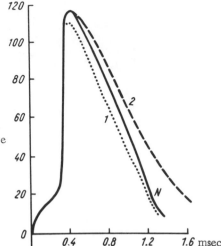

Fig. 47. Changes in shape of the calculated action potential with a decrease and increase in reactivity of the potassium system. Action potentials calculated; N) for standard values of the permeability constants; 1) for a decrease of 30 mV in φ_K; 2) for an increase of 20 mV in φ_K.

If this decrease is small (under 10 mV) the increase in I_K merely shortens the descending phase of the action potential, while leaving the amplitude and steepness of rise of the spike virtually unchanged (Fig. 46b, Fig. 47).

A further decrease in φ_K ($\Delta\varphi_K = -20$ and -30 mV) not only accelerates repolarization of the membrane, but also causes a moderate decrease in the maximum rate of rise of the action potential and a small decrease in its amplitude (Fig. 46b, Fig. 47).

The reason for these changes in the action potential is made clear by analysis of the effect of changes in φ_K.

It is clear from Fig. 48a that with a decrease in φ_K there is a sharp increase in I_K, as the result of which the net inward I_i is reduced in the ascending phase of the action potential. Correspondingly, the outward I_i is increased in the descending phase of the spike, thus accelerating repolarization of the membrane.

Characteristic changes also take place in the inward I_{Na} under these conditions: its second maximum increases to such a degree that it becomes greater than the first. The notch between the two maxima of I_{Na} is reduced both in depth and in duration.

All these effects are due to an increase in the rate of repolarization of the membrane: a rapid fall in V leads to an increase in

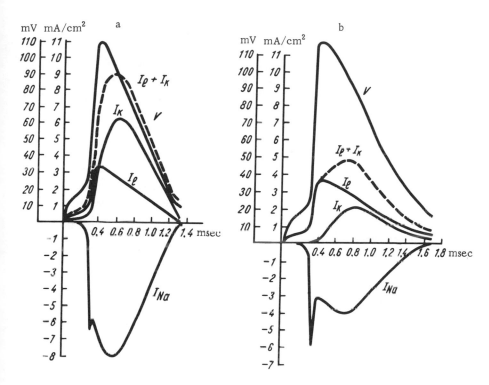

Fig. 48. Kinetics of ionic currents calculated for action potential generation in a node of Ranvier during an increase (a) and decrease (b) in reactivity of the potassium system. a) $\Delta\varphi_K = -20$ mV; b) $\Delta\varphi_K = 20$ mV, stimulation by steady current.

the electrochemical potential at a time when P_{Na} is still quite high, and I_{Na} is therefore strongly increased.

The most unexpected result of these calculations is the relative constancy of the critical potential during a marked increase (by 20–30 mV) in the reactivity of the potassium system. Essentially such a change in φ_K leads to a significant increase in P_K and I_K, not only during depolarization of the membrane, but also at rest.

For instance, with a decrease in φ_K by 20 mV ($\Delta\varphi_K = -20$ mV) the potassium permeability of the resting membrane is increased by about 57 times (from 0.86×10^{-3} to 48.7×10^{-3} cm/sec), while the potassium current is increased by 74 times (from 1.21×10^{-3} to 89.9×10^{-3} mA/cm^2). Nevertheless, the threshold potential V_t remains virtually unchanged.

It likewise is not increased when φ_K is further reduced by 30 mV, although in this case P_K rises to 143×10^{-3} cm/sec and I_K to 206×10^{-3} mA/cm^2, i.e., by about 166 and 173 times respectively. These results indicate that the magnitude of the threshold and critical potentials respectively of the node of Ranvier is determined almost entirely by the state of the sodium permeability system.

Unlike the threshold potential, the threshold strength of the current rises sharply in response to a decrease of 30 mV in φ_K (page 168). The reason for this phenomenon is that when P_K is high the net resistance of the membrane falls significantly, so that a stronger current must be applied in order to obtain a given change in potential (IR).

The results just examined are important for a correct understanding of the contribution made by an increase in potassium permeability to changes in the electrical activity of the node of Ranvier during such phenomena as cathodic depression, accommodation, and adaptation. These matters will be discussed in more detail in Chapter VII.

Effects of a Decrease in Reactivity of the Potassium

Activating System

As Fig. 46b shows, an increase of 10–30 mV in φ_K has virtually no effect on the magnitudes of the threshold potential and amplitude or of the maximum steepness of rise of the action potential. Only the duration of the descending phase of the spike is significantly altered (Fig. 47), and the greater the value of φ_K the greater the increase in its duration.

In order to understand these effects, changes in P_K and I_K must be examined during action potential generation.

The potassium permeability of the membrane rises during depolarization much more slowly than the sodium permeability (Fig. 30a), so that in the initial period of development of the action potential the outward I_K is still very weak (Fig. 30b), and its contribution to the net ionic current I_i is not significant.

It will be perfectly clear that a further decrease in the strength of I_K, produced by lowering the reactivity of the potassium system

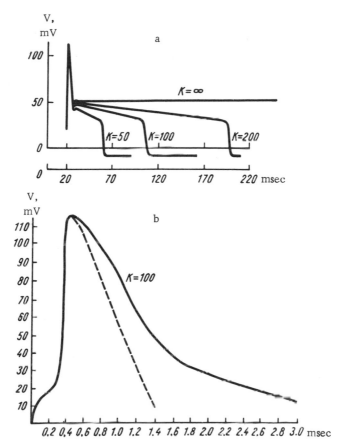

Fig. 49. Effect of time constant of potassium activation τ_n on shape of the calculated action potential in the squid giant axon (a) and node of Ranvier (b). In (a), K indicates the number of times by which τ_n is increased, temperature 6.3°C (George and Johnson, 1961); b) broken line shows descending phase of action potential calculated for standard values of τ_n; continuous line shows action potential for a 100-fold increase in τ_n [by changing the values of the constants A in Eqs. (47) and (48)]; temperature 20°C.

(Fig. 48b), can have no appreciable effect either on the threshold potential or on the steepness of the ascending phase and the amplitude of the action potential.

The principal role of the potassium current is to accelerate repolarization. With an increase in φ_K, the rate at which I_K rises to its maximum, which it attains during the action potential, is reduced, so that the descending phase of the spike becomes more prolonged.

It is an interesting fact that weakening of the outward potassium current leads to changes in the inward current of sodium ions.

With standard values of the constants the second maximum of I_{Na} has approximately the same amplitude as the first (Fig. 30b). In the case of an increase in $\Delta\varphi_K$, on the other hand, the second maximum of I_{Na} is considerably reduced in amplitude (Fig. 48b).

This change in I_{Na} is the direct result of the slowing of repolarization of the membrane, for the longer V remains at a high level, the more strongly the sodium system will be inactivated when the substantial increase in the electrochemical potential occurs during repolarization (see pages 99-102). The functional role of these changes in ionic currents is discussed in Chapter VIII.

Time Constant of Potassium Activation (τ_n)

The time constant $\tau_n = 1/(\alpha_n + \beta_n)$ determines the rate of changes of P_K and I_K during stepwise depolarization or repolarization of the membrane. FitzHugh (1960) and George and Johnson (1961), using the squid giant axon as their model, showed that a 20- to 70-fold increase in τ_n leads to the appearance of a long plateau in the descending phase of the action potential (Fig. 49a).

In the node of Ranvier this effect (the appearance of a plateau) does not arise even in response to a 100-fold increase in τ_n. The crest of the spike simply becomes rounded and the repolarization phase is protracted (Fig. 49b).

These differences in the effects of prolonging τ_n in the node of Ranvier and in the squid giant axon are explained as follows. In the squid giant axon the outward ionic current causing repolarization of the membrane is almost entirely determined by I_K, since the leak conductance g_l for this excitable structure is very small ($\sim$0.3 mmho/cm^2) (Fig. 50).

In the medullated nerve fiber the leak conductance is about 30 mmho/cm^2 and the contribution of I_l to the net outward current is very large (Fig. 30b). The node membrane can therefore be repolarized even if the potassium current is practically completely blocked. However, in such a case repolarization takes place more

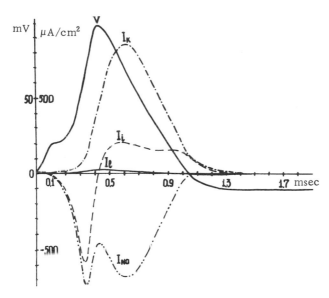

Fig. 50. Kinetics of ionic currents during action potential generation in the squid giant axon. Calculated for stimulation of the axon by a short electrical stimulus, 20°C.

slowly and it does not begin until the value of I_{Na}, as it falls because of inactivation, becomes less than the outward leak current. Later, besides inactivation there is regenerative repolarization, resulting from reversal of the activation process (page 102) as the membrane potential draws closer to the resting level.

The results relating to the effect of changes in $\overline{P}_K$, the reactivity of the potassium system, and τ_n on the parameters of the action potential examined above also help to explain the mechanism of the changes in the electrical activity of the nerve fiber produced by agents which inhibit the potassium activation process.

The following such agents are noteworthy:

1. Polyvalent cations Ca^{++}, Ni^{++}, Ba^{++}, and La^{+++}. These all reduce the reactivity of the potassium system, i.e., they increase φ_K (Frankenhaeuser and Hodgkin, 1957; Frankenhaeuser, 1957; Blaustein and Goldman, 1966b, 1968; Takata et al., 1966; Khodorov and Peganov, 1969; Mozhaeva and Naumov, 1970), thereby creating a tendency for the descending phase of the spike to be prolonged. However, this effect is complicated by the strong in-

fluence of these ions on the characteristics of the sodium system (changes of φ_{Na}, h_∞, etc.). For instance, an excess of Ca^{++} ions increases φ_{Na} and shortens τ_m (Frankenhaeuser and Hodgkin, 1957), as the result of which the regenerative closing of the sodium channels (see page 125) and, correspondingly, repolarization of the membrane, are accelerated (Ulbricht, 1964). During the action of Ni^{++}, Co^{++}, and Cd^{++} ions the increase in φ_K and φ_{Na} is combined with a simultaneous increase in τ_m, τ_n, and τ_h (Dodge, 1961; Meves and Weymann, 1963). As a result the duration of the descending phase of the action potential is greatly increased. A similar action is shown by La^{+++} ions, but at the same time they reduce the values of $\overline{P}_{Na}$ and $\overline{P}_K$ (see pages 286-287).

2. Local anesthetics, such as procaine, cocaine, xylocaine, etc., lower $\overline{g}_K$ evidently in all excitable structures, thus lengthening the descending phase of the action potential (Shanes et al., 1959). However, the predominant feature of the effect of local anesthetics is inactivation (a decrease in h) and a decrease in the maximum sodium permeability, leading to suppression of spike generation (Taylor, 1959; Khodorov and Belyaev, 1965b, 1967; Hille, 1966, 1970). The time constants τ_m and τ_n are unchanged by the action of local anesthetics in the node of Ranvier (Hille, 1966), while in the squid giant axon τ_m and τ_n are slightly increased (Taylor, 1959).

3. Tetraethylammonium (TEA) has a specific inhibitory effect on the activation of potassium permeability, as has already been mentioned. If tetraethylammonium is injected inside the giant axon, it leads to a very considerable prolongation of the descending phase of the action potential, accompanied by the formation of a characteristic plateau. These "prolonged" action potentials are very similar in shape to those obtained by George and Johnson (1961) in the giant axon in response to a severalfold increase in the time constant of potassium activation (Fig. 49a).

However, the investigations of Armstrong and Binstock (1965) showed that injection of tetraethylammonium inside the axon does not affect τ_n, but the maximum potassium conductance $\overline{g}_K$. The most interesting fact is that potassium conductance falls only for the outward current; the inward I_K is virtually unchanged under these conditions.

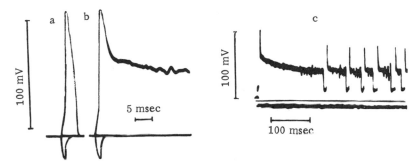

Fig. 51. Prolongation of action potential after combined application of tetraethylammonium (TEA) and Ni^{++} ions: a) action potential of single node of Ranvier in Ringer's solution containing 1 mM $NiCl_2$; b) the same after addition of
10 mM TEA to the solution. Short stimuli applied; c) prolonged action potential
changing into repetitive response. Node in Ringer's solution containing 1 mM
$NiCl_2$ and 10 mM TEA. Stimulation by steady current.

A somewhat different picture is observed during the action of
tetraethylammonium on nodes of Ranvier. In this case tetraethylammonium is effective also when applied externally,* and the value
of $\overline{P}_K$ falls not only for the outward, but also for the inward potassium current; τ_{ll} is increased in duration under these circumstances
(Koppenhöfer, 1967). Despite this fact, the prolongation of the descending phase of the spike in the node of Ranvier under the influence of tetraethylammonium is comparatively moderate in degree: it is approximately doubled in sensory fibers and increased
by three times in motor fibers (Schmidt and Stämpfli, 1966) (Fig.
65). The reason for this is that in medullated nerve fibers, as
already mentioned, g_l is high, and the outward I_l plays as important
a role as I_K in the mechanism of membrane repolarization.

The appearance of "long" action potentials in nodes of Ranvier
under the influence of tetraethylammonium was observed only when
its action was combined with that of nickel ions. In this case, as
Fig. 51 shows, a plateau is formed and, in some cases it lasts
longer than 100 msec. The mechanism of this combined action of
tetraethylammonium and nickel has not yet been analyzed (Khodorov
and Belyaev, 1965a, 1966a).

* TEA, applied to the inside of the node membrane through the cut end of the fiber,
  also reduces I_K (Koppenhöfer and Vogel, 1969).

Assessment of the Character of Changes

in Ionic Permeability of the Membrane

from Changes in the Action Potential

It is now clear that the most adequate method of analysis of
the properties of an excitable membrane is the voltage-clamp method. However, the use of this method is beset by considerable technical difficulties, and it is by no means always practicable. The
question therefore naturally arises: to what extent can the character of changes in the ionic permeability of an excitable membrane
be assessed indirectly, from changes in the parameters of the membrane action potential? Some general conclusions on this matter
can be drawn from the above analysis of the relationship between
the threshold potential, rate of rise, amplitude, and duration of the
action potential of the node of Ranvier and the various constants
of sodium and potassium permeability of the membrane:

I. Let us first consider changes in the rising phase of the
action potential. Calculations show that the maximum rate of rise
of the action potential ($\dot{V}_{max}$) of the nerve fiber is directly dependent on the first maximum I_{Na}. If I_{Na} is weakened, $\dot{V}_{max}$ falls,
and vice versa.

However, changes in $\dot{V}_{max}$ themselves are insufficient to solve
the problem of the changes in the sodium system of the membrane
responsible for the increase or decrease in I_{Na}.

We saw above that weakening of I_{Na} can be due to various
causes:

1. A decrease in the reactivity of P_{Na} (a shift of the P_{Na}
    versus V curve toward higher values of V, i.e., an in-
    crease in φ_{Na});
2. An increase in the time constant of sodium activation (τ_m);
3. An increase in the steady-state inactivation (a decrease
    in h_∞);
4. A decrease in the maximum sodium permeability $\bar{P}_{Na}$.

In order to differentiate between these changes, changes in
$\dot{V}_{max}$ must be compared with the corresponding changes in threshold potential and amplitude of the action potential.

A moderate decrease in $\dot{V}_{max}$, accompanied by a simultaneous
large increase in the critical potential but with the amplitude of

the action potential unchanged or only slightly reduced, is characteristic of a decrease in the reactivity of the sodium system (see
Figs. 42b and 43b). As was mentioned above (page 126), it is observed experimentally when the Ca^{++} ion concentration in the medium is increased under conditions in which sodium inactivation of
the nerve fiber in its original state is weak or has been specially
reduced by the action of a hyperpolarizing current.

Somewhat different relationships between $\dot{V}_{max}$, V_t, and the
amplitude of the action potential are found when $\dot{V}_{max}$ is reduced
as the result of an increase in the time constant of sodium activation τ_m.

In this case a marked decrease in $\dot{V}_{max}$ is accompanied by a
comparatively small increase in the threshold potential, while the
amplitude of the action potential remains unchanged (Fig. 11).

Finally, if h_∞ or the maximum sodium permeability is decreased, besides the decrease in $\dot{V}_{max}$, there is also a decrease
in amplitude of the action potential, which may actually disappear
completely, and an increase in the threshold potential (Figs. 31-35).

These changes are characteristic of the effects of procaine,
tetrodotoxin, urethane, and other agents inhibiting action potential
generation.

The rate of rise of the action potential is only slightly sensitive to changes in potassium permeability of the membrane.
Even if $\overline{P}_K$ and I_K are considerably increased, there is only a
moderate decrease in $\dot{V}_{max}$, accompanied by a small decrease in
spike amplitude. The threshold potential remains unchanged (see
Figs. 46b and 17).

The diagnostic features of this state of the membrane are:
shortening of the duration of the spike and a considerable increase
in the threshold strength of the current (page 168) which results
from a decrease in the resistance of the excitable membrane while
the threshold potential remains virtually unchanged in magnitude.

Changes in the amplitude and shape of the action potential in the
case of a decrease in the rate of rise resulting from various changes
in P_{Na} and P_K can thus be successfully differentiated.

Let us now analyze changes in the action potential accompanied
by an increase in the rate of rise.

Calculations have shown that the increase in $\dot{V}_{max}$ can be due to (1) an increase in the sensitivity of P_{Na} to depolarization (a decrease in φ_{Na}), or (2) a reduction of sodium inactivation, i.e., an increase in h_∞ if it was low in the original state of the preparation.

In order to differentiate between these states of the membrane, changes in $\dot{V}_{max}$ must be compared with changes in V_t and in the amplitude of the action potential.

In the first case, i.e., if the sensitivity of the sodium system is increased, the increase in $\dot{V}_{max}$ is combined with a decrease in the critical level of depolarization: the amplitude of the action potential is virtually unchanged under these conditions, while the duration of the descending phase of the spike is increased (see Figs. 42b and 43c).

In the second case, hyperpolarization will remove the inactivation, and the initially depressed $\dot{V}_{max}$ and amplitude of the action potential and the increased threshold potential are restored to their initial values (Fig. 37a, Fig. 39a,c,d).

II. A more complex analysis of the causes of changes in the duration of the descending phase of the action potential. This duration can be increased by changes in either the sodium or the potassium permeability of the membrane.

An increase in the duration of the repolarization phase can be due to:

1. A decrease in reactivity of the potassium system (an increase in φ_K) or a decrease in the maximum potassium permeability $\bar{P}_K$;
2. An increase in the time constant of potassium activation τ_n;
3. A decrease in the initial level of sodium inactivation (an increase in h_∞), if it had hitherto been well marked;
4. An increase in the time constant of sodium inactivation τ_h;
5. An increase in the reactivity of the sodium system (a decrease in φ_{Na}).

The first two changes are characterized by an increase in length of the descending phase of the action potential, while the threshold, steepness of rise, and amplitude of the action potential retain their original values (Figs. 46b, 47, 48b, 49).

Removal of the initial inactivation can bring about restoration of the previously shortened descending phase of the spike, but can in no way prolong the phase of repolarization beyond the normal limits. Another sign of this change in the state of the membrane is a simultaneous increase in $\dot{V}_{max}$ and the amplitude of the spike to their original values (Fig. 37a).

With an increase in τ_h, the repolarization process is very protracted, and a more or less conspicuous plateau is formed; all the other parameters of the action potential are virtually unchanged (Fig. 36).

Finally, prolongation of the descending phase of the action potential with an increase in the reactivity of the sodium system can be identified from the simultaneous decrease in the threshold potential and increase in $\dot{V}_{max}$ (Fig. 43c).

III. We must now examine the possible causes of a decrease in the duration of the descending phase of the action potential. These are as follows:

1. An increase in reactivity of the potassium system (a decrease in φ_K);
2. Shortening of the time constant of potassium activation τ_u;
3. An increase of the initial inactivation of sodium permeability (a decrease in h_m);
4. A decrease in the time constant of sodium inactivation τ_h;
5. A decrease in the reactivity of the sodium system (an increase in φ_{Na}).

Since the characteristics of each of these states have already been examined, it is unnecessary to mention the methods of their differential diagnosis again.

Clearly, therefore, the results obtained with the mathematical model of the excitable membrane can be of great importance to the identification of changes in the threshold, amplitude, and shape of the action potential arising in response to the action of various factors on the nerve fiber.

The chief difficulty in such an analysis is that most factors do not cause changes in any single characteristic but simultaneously in several characteristics of the ionic permeability of the membrane, so that the final effect, i.e., the change in the parameters of the action potential, is very complex.

It was shown above, for example, during the analysis of the effects of Ca^{++} ions, that changes in the rate constants of sodium activation and sodium inactivation, taking place in the same direction, create a tendency toward directly opposite changes in the parameters of the action potential. As a result, with an increase in $[Ca]_0$, depending on the initial level of inactivation of the membrane, the rate of rise and the amplitude of the action potential will either increase, decrease, or remain virtually unchanged (see pages 126-128).

To form a correct estimate of the character of the changes in ionic permeability of the membrane it is therefore necessary to make a very careful analysis of the action potential, taking into account all the results obtained in investigations on the appropriate mathematical model. Unfortunately an investigation of this type has so far been undertaken only on a model of the membrane of the medullated nerve fiber.

The result thus obtained is regarded as hopeful and indicative of the desirability of extending this type of analysis to models of the membrane of other excitable structures. In particular, the study of the relationship between the parameters of the action potential and the constants of ionic permeability of the membrane of the Purkinje fibers and cells of the myocardium could be of great theoretical as well as practical importance.

Chapter VI

Analysis of Threshold Conditions of Stimulation and the Relationship Between the Threshold Current and Threshold Potential

The threshold potential is an important parameter of excitability of the cell. However, its measurement still does not enable the investigator to answer the question whether a stimulus applied to a membrane will be adequate to induce the generation of an action potential. In order to answer this question, as well as the threshold potential, it is also necessary to determine the threshold current, i.e., the smallest strength (density) of a current passing through the membrane which will be adequate to produce its critical depolarization.

By contrast with the threshold potential, the threshold current (I_t) is dependent not only on the active, but also on the passive electrical characteristics of the excitable membrane and of the cell as a whole; it also depends on the duration of action of the stimulus. For this reason, the study of factors determining the magnitude of the threshold current deserves special attention.

In this chapter we shall examine this question for the case of action potential generation in a homogeneous area of an excitable membrane. The influence of the cable properties of the nerve fiber on the threshold conditions of stimulation and characteristic of the action potential will be examined in Chapter VIII.

The Strength versus Duration Curve

This is the term usually given to the curve showing how the

threshold strength of the current varies with the duration of its action on the excitable tissue.

This curve was first obtained by Hoorweg in 1892 when investigating the excitability of human skeletal muscles by means of condenser discharges of different durations. Later the same relationship was established by Weiss (1901), who used square pulses of steady current to stimulate a frog nerve-muscle preparation. To describe this curve quantitatively, both these workers proposed the formula

$$i = \frac{a}{t} + b,$$

(49)

where i is the threshold strength of the current; t is the duration of the stimulus; and a and b are constants.

For very long stimuli, i.e., when t → ∞, a/t → 0, and i = b. Consequently, the constant b corresponds to the threshold strength of a current of long duration, i.e., to the rheobase.

For very short stimuli, i.e., when t → 0, the term a/t becomes much greater than b. Disregarding b, we obtain

$$a = it.$$

(50)

This means that the constant a characterizes the threshold quantity of electricity contained in pulses of current of very short duration. Later investigations by Lapique (1907-1926, 1936) and other workers showed that strength versus duration curves are similar in character for widely different excitable structures, such as nerves and skeletal muscles of vertebrates and invertebrates, heart muscle, smooth muscles, filaments of algae, protozoa, and so on. The differences discovered in these investigations were concerned mainly with the position of the curve as a whole relative to the time axis.

Since the determination of a large number of points on the strength versus duration curve is a laborious procedure and one which is bound to be harmful to the excitable structure (especially if the excitable structure is treated in certain ways), Lapique pro-

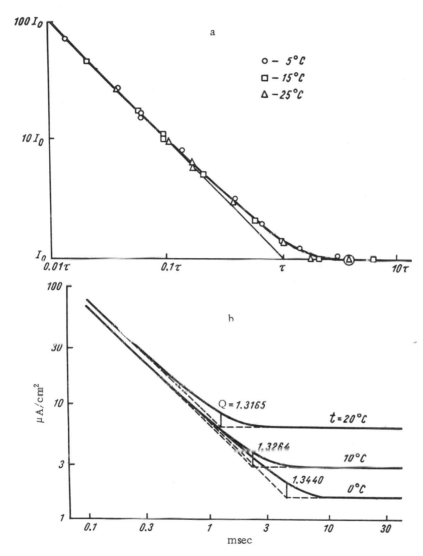

Fig. 52. Strength versus duration curves of a squid giant axon: a) experimental data. Ordinate, density of stimulating current (I) in rheobase units (I_0). Abscissa, duration of stimulus in units of time constant of stimulation τ_s; both axes logarithmic (Guttmann and Barnhill, 1966); b) strength versus duration curve calculated for a squid giant axon at different temperatures. Duration of stimuli in msec; density of stimulating current in $\mu A/cm^2$ (FitzHugh, 1966).

posed measuring only two of its parameters to characterize this curve as a whole: the rheobase (b) and the chronaxie (t_{chr}).

The term chronaxie was used to define the minimum duration of a stimulus with a strength equal to double that of the rheobase. This point was not chosen accidentally.

It follows from Eq. (49) that, when i = 2b,

$$t_{chr} = \frac{a}{b} \cdot \qquad\qquad\qquad (51)$$

The chronaxie is thus defined as the ratio between the quantity of electricity contained in pulses of current of short duration and the rheobase.

The experimental verification of this hypothesis was for a long time attended with great difficulty, for the experiments were not conducted on single fibers, but on whole nerve trunks. Furthermore, no methods of accurately measuring the administering pulses of current with a duration measured in microseconds were then available (Rosenberg, 1935).

Nevertheless, a number of investigators succeeded in overcoming these difficulties in experiments on nerves of both cold-blooded (frog, cockroach, crustaceans), and on warm-blooded (cat, dog) animals, and they showed that if very short pulses of current are used the threshold quantity of electricity is approximately constant (Rosenberg, 1935; Hill, 1936b).

This fact became very obvious after experiments had been performed in which potentials were recorded from isolated nerve fibers.

A strength versus duration curve obtained by Guttmann and Barnhill (1966) on a squid giant axon and plotted in a system of double logarithmic coordinates is shown in Fig. 52a.

It will be seen that this curve consists of two branches. One branch becomes a straight line with slope of −1 in the region of short values of t. This straight line corresponds to the constant value Q = it ("the law of constancy of the quantity of electricity"). The second branch of the curve becomes a horizontal line at long durations of the stimuli. It reflects the independence of the threshold strength of the current (the rheobase) of time as t → ∞.

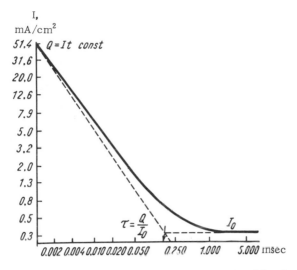

Fig. 53. Strength versus duration curve calculated for the membrane of a single node of Ranvier. Abscissa, duration of stimulus in msec, ordinate, stimulating current in mA/cm^2. Both scales logarithmic (Grilikhes, Rusanov, and Khodorov, 1967).

These two straight lines intersect at the point $t = \tau_s$ which, according to Laplique, must correspond to the chronaxie, for at this point the product of rheobase and time is equal to the quantity of electricity contained in short pulses of current. Since the chronaxie is the minimum time necessary for stimulation of the fiber by a current equal to twice the rheobase, the value of I/I_0 at the point τ_s must be 2.0.

However, the experiments of Guttmann and Barnhill (1966) showed that in the squid giant axon I/I_0 at the moment $t = \tau_s$ is not 2.0, but only 1.38. FitzHugh (1966) designated this value σ and proposed using it, together with the parameters of the rheobase I_0, the threshold quantity of electricity Q, and the time constant of stimulation $\tau_s = Q/I_0$, as adequate characteristics of the strength versus duration curve. The chronaxie parameter and the equation of Hoorweg and Weiss (49) from which it was obtained, in Fitz-Hugh's opinion, cannot be used to describe the true strength versus duration curve.

For the squid giant axon, as was mentioned above, the mean value of σ is 1.38 (Guttmann and Barnhill, 1966). According to these workers, the rheobase of the giant axon at 20°C is 12 μA/cm², while $Q \sim 1.5 \times 10^{-8}$ C/cm². Variation of the temperature between 5 and 35°C has little effect on the value of Q, but it changes the rheobase substantially: if the temperature rises 10°, the rheobase rises 2.35 times.

Investigations on squid giant axons showed the inadequacy of other equations suggested for the description of the strength versus duration curve (Lapique, 1907; Blair, 1932; Hill, 1936b). The best agreement with the experimental data was obtained by Cole, Antoziewicz, and Rabinowitz (1955) and FitzHugh (1966) by solution of the equations of Hodgkin and Huxley (1952d).

For instance, FitzHugh (1966) found that σ for the strength versus duration curve calculated with the aid of these equations at temperatures of between 0 and 20°C is 1.32–1.34, which is practically indistinguishable from the experimentally obtained value (1.38).

Calculation of the strength versus duration curve for nodes of Ranvier were made by Grilikhes, Rusanov, and Khodorov (1967).

It will be seen in Fig. 53 that this curve is very similar in its character to the strength versus duration curve obtained with the squid giant axon model, but differs from it in its position relative to the voltage and time axes. For instance, the value of the rheobase (I_0) for nodes of Ranvier (0.343 mA/cm²) is about 20 times greater than the rheobase of the giant axon, while the time constant of stimulation τ_s (about 0.15 msec) is 10 times shorter than τ_s for the giant axon.

The differences between the values of the rheobase are understandable: they are attributable mainly to the fact that the resistance of the squid giant axon membrane is much higher than the resistance of the node of Ranvier (1000 Ω/cm² and 10–30 Ω/cm² respectively).

In order to understand the reason for the differences between the values of the time constant τ_s, the question of the nature of the utilization time must be examined.

Character of Critical Depolarization

of the Membrane during the Action of

Pulses of Current of Different Duration.

Nature of the Utilization Time

The term "utilization time" ("Nutzzeit," "duré utile") in physiology is taken to mean the shortest time interval during which an electric current of given strength must act in order to evoke the generation of a spike. The utilization time of a stimulus of rheobase strength is called the "principal utilization time" ("Hauptnutzzeit") (Gildemeister, 1913).

Before the discovery of the local response, i.e., before 1937–1938, most investigators considered that critical depolarization of the membrane takes place only through passive (electronic) changes of membrane potential. The utilization time was therefore regarded as the necessary and adequate time for charging the membrane up to the threshold level. Since in a uniform membrane the rate of the potential increase for a given strength of current is determined by the time constant $\tau = R_m C_m$, the latter was identified with the time constant of stimulation.

This point of view of the mechanism of action of the electric current was expressed in 1906 by Chagovets ("the condenser theory of excitation") and subsequently by Lapique (1907).

In the paper cited, Lapique proposed the following equation to describe the strength versus duration curve:

$$I/I_0 = \frac{1}{1 - e^{-t/R_m C_m}},$$
(52)

where I_0 is the rheobase, I the threshold strength of current of duration t, and $R_m C_m$ the time constant of the membrane. The chronaxie equals $0.69\, R_m C_m$.

Similar concepts were subsequently developed by Blair (1932), Ebbecke (1926, 1933), Cremer (1900, 1932), Rosenberg (1935), Schäfer (1935, 1936), and Hill (1936).

All these workers identified the utilization time of stimulation with the latent period of the action potential and considered that the action potential is generated when the electrotonic potential difference reaches the threshold value.

The first doubts about the validity of these assumptions were raised by the investigations of Blair and Erlanger (1936). In experiments on a frog phalangeal preparation they found that during stimulation of single nerve fibers by very short stimuli (of the order of 0.04 msec) the latent period of the action potential may be much longer (by 0.1–0.4 msec) than the duration of the pulse of current itself.

The curve of latent period as a function of stimulus duration is of the same character as the strength versus duration curve, from which it differs only in its position relative to the abscissa. The utilization time and the latent period of the action potential coincide in duration only for a current of rheobase strength. As the strength of the stimuli increases, the time interval between the break of the current and the beginning of the spike increases rapidly at first, and then more slowly, and for short pulses it becomes approximately constant.

These facts, subsequently confirmed by Tasaki and Fujita (1948) on a single Ranvier node, contradicted the view of the purely passive character of critical depolarization of the membrane. Blair and Erlanger (1936) therefore attempted to discover whether some sort of intermediate process of excitation develops in the time between the end of the stimulating current and the beginning of the action potential.

As we now know, this guess proved correct, although the methods available for recording potentials at that time were not sufficiently refined to allow Blair and Erlanger to detect the local response preceding the spike.

They therefore postulated that the discrepancy between the values of the useful time and latent period of stimulation is merely apparent and is due to the effect of the capacitances and resistances of the nerve trunk membranes on the rise of potential on the excitable membrane.

That this assumption was incorrect became clear after the work of Hodgkin (1938), Stämpfli and Huxley (1951), and Castillo and Stark (1952), who used isolated nerve fibers and showed that in response to short stimuli a spike in fact is generated not as the stimulating current passes through the membrane, but only after

the current is broken, when the gradually increasing local response reaches the critical level of depolarization (Fig. 16b).

The reason why this is so is that during the short period while
the stimulus lasts the sodium permeability and, correspondingly,
the inward sodium current do not have have time to increase sufficiently, and the further increase responsible for generating the
local response and action potential takes place only after the end
of the stimulus, as it were on the traces of the waning electronic potential.

Subsequent calculations made by the author in conjunction
with Grilikhes, Rusanov, and Korotkov (see Khodorov, 1967)
gave a much clearer idea of the kinetics of the ionic currents
arising during stimulation of a nerve fiber with s h o r t stimuli.

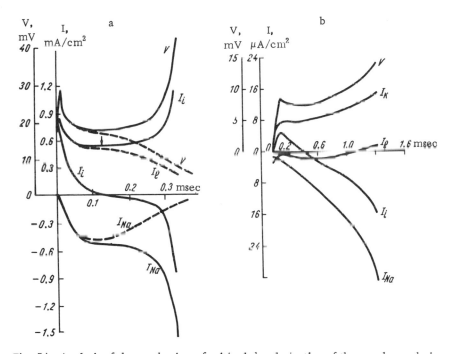

Fig. 54. Analysis of the mechanism of critical depolarization of the membrane during
the action of stimuli of short duration: a) model of membrane of node of Ranvier;
temperature, 20°C, square pulses, 0.1 msec in duration, of threshold (6.23 mA/cm²,
continuous lines) and subthreshold (6.20 mA/cm², broken lines) density. Abscissa,
time in msec; ordinate, V, in mV; I, in mA/cm²; b) model of membrane of squid
giant axon (temperature 6.3°C). Pulse duration 0.065 msec. Stimulation at suprathreshold strength (I_m = 100 µA/cm²). Abscissa, time in msec; ordinate: V, in mV;
I, in µA/cm².

The initial changes in membrane potential and ionic currents calculated for stimulation of a node of Ranvier by short (0.01 msec) pulses of current of threshold and very near-threshold strength are illustrated in Fig. 54a.

It will be clear from this figure that in both cases during the time of action of the stimulus the potential V rose to a certain value V', and after the current had been broken it began to fall gradually. The differences between the initial rise in response to threshold and subthreshold stimuli are very small (29.85 mV and 29.79 mV respectively), but nevertheless they were sufficient to cause a local response to rise in the first case, developing into an action potential (only its initial part is shown in the figure), whereas in the second case only the local response was generated. The reason for these differences becomes clear when the kinetics of the ionic currents is examined.

In the initial period of the changes in membrane potential the leak current I_l evidently reaches its maximum, and changes in I_l repeat changes in V precisely. This is because the leak conductance g_l is practically independent of V (see page 67), so that linear relationships hold good all the time between I_l and V. The inward sodium current during the action of the stimulus does not have sufficient time to increase substantially ($\tau_m \gg RC$), and for this reason the net ionic current I_i is virtually identical with I_l at this time.

After the end of the stimulus, V and I_l begin to fall. If no increase in the inward current I_{Na}, i.e., the current depolarizing the membrane, took place during this time, the membrane potential would return to its initial level at a rate determined by $R_m C_m$ of the membrane.

In fact, however, I_{Na} does increase, and this delays the fall in V. The delay in repolarization, however, facilitates a further increase in I_{Na} which, in turn, delays the fall in I_{Na} still further. As a result of this process, 0.11 msec after the end of the stimulus I_{Na} becomes equal to the sum of I_l (0.529 mA/cm^2) and I_K (0.004 mA/cm^2). At this moment the decrease in V has completely ceased ($\dot{V} = 0$), while I_{Na} continues to increase. Now I_{Na} becomes greater than $I_l + I_K$, and the total ionic current I_i becomes inward in direction, i.e., it depolarizes the membrane, and generation of the action potential begins.

With a current of subthreshold strength (Fig. 54a, broken lines) toward the end of action of the stimulus, V has increased by a slightly smaller amount than in response to threshold stimulation. As a result of this, the increase in sodium permeability was too small to enable I_{Na} to reach the combined value of $I_l + I_K$. The closest which I_{Na} could reach to this value occurred 0.1063 msec after the end of the stimulus, when V = 16.88 mV. At this moment I_{Na} reached 0.482 mA/cm² and was only 0.083 mA/cm² less than the value of $I_l + I_K$.

Since the increase in I_{Na} in the case under examination did not stop the decrease in V, this decrease in turn caused a decrease in I_{Na}. The outward I_K under these circumstances began

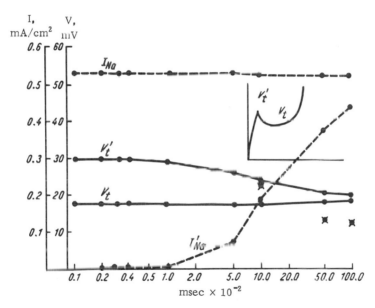

Fig. 55. Threshold conditions for stimuli of different durations. Membrane of Ranvier node used as model. V_t' represents the potential at the end of stimulus of the given duration; I_{Na} the sodium current at that moment; V_t the threshold potential; I_{Na} the sodium current at the moment of reaching the threshold potential. Thresholds were calculated with an accuracy of 0.1%. Abscissa, duration of stimuli in msec (logarithmic scale); ordinate, change of potential (V) in mV and density of sodium current in mA/cm². Crosses show what would be the value of the initial change in potential V_t' at the end of the stimulus had this change been purely passive in character; 20°C (Grilikhes, Rusanov, and Khodorov, 1967).

to rise, and this accelerated repolarization of the membrane still more.

A similar picture of the relationships between the ionic currents is also found in the squid giant axon (Fig. 54b). The only difference is that in this excitable structure, just as evidently in most other structures except medullated nerve fibers, the potassium conductance of the resting membrane is much greater than the leak conductance (see page 136), so that the leading role in the process of repolarization of the membrane is played not by the leak current, which, as is clear from Fig. 54b, remains inward in direction in the initial period of depolarization (since at rest $V - V_l = -10$ mV), but by the outward potassium current. Growth of the local response into the spike thus takes place at the moment when $I_{Na} + I_l$ becomes equal to I_K, i.e., when $I_i = 0$.

In the course of investigation of the threshold conditions of stimulation on the model of the membrane of the node of Ranvier, the duration of the stimulus (t) was varied from 0.001 to 1 msec (Fig. 55).

Calculations showed that when t < 0.5 msec the threshold change in membrane potential V_t (at which I_i and $\dot{V}$ become equal to zero) is about constant in value: 17.4-17.5 mV. Only at longer durations of the current does V_t increase slightly (at t = 1 msec to 18.3 mV) as the result of inactivation of the sodium system developing during this time.

However, it must be emphasized that when V_t is determined, diligent attempts must be made to find the threshold strengths of current. In the writer's calculations it was determined with an accuracy of not less than 0.1-0.5%, for even if this strength was only a little above threshold the increase in I_{Na} took place much more rapidly, and equality between the inward and outward currents took place at a higher value of V.

Analysis of the relationship between the initial change of potential V' and the duration of action of the threshold stimulus deserves special attention.

Calculations showed that for currents of very short duration (of the order of 0.001-0.003 msec) the initial change in potential V' is approximately equal (about 29 mV) to the value calculated by

Eq. (8). This means that during these time intervals the initial change in potential is almost entirely due to passive depolarization of the membrane. The threshold quantity of electricity contained in these short pulses is almost constant (5.14×10^{-8} C/cm^2).

With an increase in the duration of the stimuli, the initial passive change in membrane potential developed during the action of the cover pulse begins to be supplemented by active changes in V associated with the increase in I_{Na}. The contribution of the active component (the local response) to the initial change in V' becomes particularly evident when the stimulus duration reaches 0.5-1 msec. For instance, with a pulse of 1 msec, if the change in potential V' had been purely passive in character, it would have been about 12 mV (see Fig. 55), whereas in fact it was 19.74 mV. Consequently, 40% of the initial change in this case is attributable to the local response.

With differences in the duration of stimuli of strictly threshold strength, sometimes purely passive, and at other times mixed (passive and active) changes in membrane potential take place during the time of their action, i.e., during the utilization time. The role of these changes in the mechanism of initial depolarization of the nerve model becomes greater with an increase in the duration of the stimulus.

The mechanism of critical depolarization of the membrane during stimulation of the nerve fiber by a steady current requires special attention. In this case, as mentioned above, action potential generation takes place during the ascending phase of the electrotonic potential, and the role of the local response and, consequently, of I_{Na} which generates it, is simply to bring depolarization of the membrane up to the critical level (Figs. 13a and 16a).

Changes in V, $\dot{V}$, and ionic currents calculated for the stimulation of the mode with a steady current of density 1.5 times the rheobase are shown in Fig. 56.

An increase in V initially caused an increase in the outward leak current, which was practically the only component of the net ionic current I_i during the first 0.05-0.08 msec. A further increase in V led to an increase in I_{Na} which, in turn, accelerated the development of depolarization. Until the rate of increase of I_{Na} became less than the rate of increase of $I_i + I_K$, the net ionic cur-

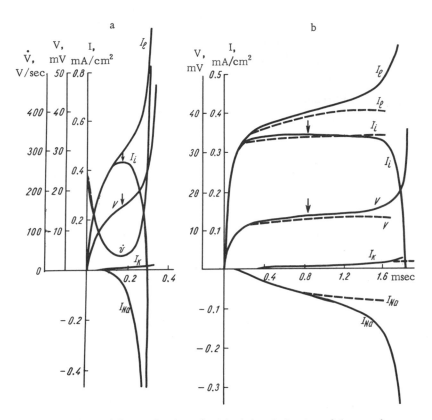

Fig. 56. Analysis of the mechanism of critical depolarization of the membrane during the action of a steady current. Model of membrane of node of Ranvier: a) above-threshold current (0.5 mA/cm²), b) threshold (0.343 mA/cm², continuous lines) and subthreshold (0.340 mA/cm²) density. Abscissa, time in msec; ordinate, $\dot{V}$ in V/sec, V in mV, and I in mA/cm². Arrows indicate times when dI_i/dt passes through zero (Khodorov, 1967).

rent I_i continued to increase, although its density was already smaller than I_l. However, 0.168 msec after the beginning of stimulation, dI_{Na}/dt became equal to, and then greater than $d(I_l + I_K)/dt$ so that the net ionic current I_i began to fall. From this moment the rate of depolarization began to rise sharply.

The calculations showed that similar changes in ionic currents also take place in response to a stimulating current of rheobase density $I_m = 0.343$ mA/cm². In this case (see Fig. 56b, continuous lines), however, the changes in V, I_l, and I_{Na} develop much more slowly, and I_i does not begin to fall until 0.95 msec after the beginning of application of the steady current. The value of V at this moment was 13.46 mV, and $I_{Na} = 0.0757$ mA/cm^2 when $I_l + I_K = 0.416$ mA/cm^2. The equality: $I_{Na} = I_l + I_K$ occurred only after 1.8 msec at V = 24.5 mV.

In response to a stimulating current of subthreshold density (0.34 mA/cm^2), as is clear from Fig. 56b (the broken lines), the rate of increase of I_{Na} is always slower than the rate of increase of $I_l + I_K$, so that I_i retains its outward direction, and it gradually increases throughout the period of action of the stimulating current. In the case under examination the slow increase in I_{Na} caused active subthreshold depolarization of the membrane (a local response), the magnitude of which reached a maximum (V = 13.45 mV) after approximately 1.5 msec, after which it began to decrease gradually under the influence of inactivation and the increase in outward current.

During stimulation with a steady current, $\dot{V}$ undergoes characteristic changes (Fig. 56a). At the moment of application of the current $\dot{V}$ increases, but later, with an increase in V, it falls and then again begins to rise. The minimum through which $\dot{V}$ passes, as these figures show, differs from zero ($\dot{V} > 0$) and corresponds to a change in direction of the changes in I_i.

These relationships between I and $\dot{V}$ can be explained as follows. It follows from Eq. (14) that

$$\frac{dV}{dt} = \frac{1}{C}(I_m - I_i). \tag{53}$$

During the action of the steady current $I_m > 0$, and it remains constant in magnitude all the time, so that on application of the current $\dot{V}$ is also positive. As I_i increases (on account of I_l), the difference $I_m - I_i$ and, correspondingly, $\dot{V}$ decrease. When I_i reaches its maximum, $\dot{V}$ is minimal; the decrease in I_i due to the increase in I_{Na}, however, leads to a new increase in $\dot{V}$. During subthreshold stimulation $\dot{V}$ also falls, but this fall is not followed by a later rise. On the contrary, it passes through zero and ultimately becomes negative.

This means that the passage of $\dot{V}$ through the minimum, at which the rate of increase of the inward sodium current becomes equal to the rate of increase of the outward current

$$\frac{dI_{Na}}{dt} = \frac{d(I_1 + I_K)}{dt},$$

is an essential condition for action potential generation. Correspondingly, the value of V at which $\dot{V}$ is minimal is the critical value.

This analysis thus shows that during stimulation of the nerve fiber with a steady current, by contrast with what happens in the case of short stimuli, the condition of generation of the action potential is not equality of the absolute values of the inward and outward ionic currents, but equality of their rates of increase: the value of V at which this condition is satisfied is a U-shaped function of the strength of the steady current (Table 2).

With a current I_m = 0.360 mA/cm², of suprathreshold strength, V_t is thus 13.01 mV, while at the rheobase value I_m = 0.343 mA/cm², V_t rises to 13.46 mV, and if the strength of the current is increased to 0.8 mA/cm², V_t rises to 20.27 mV.

In a special analysis of the threshold conditions of stimulation, Noble and Stein (1966) described the threshold potential as the

TABLE 2. Threshold Potential as a Function of
Strength of the Steady Current (Khodorov, 1967)

Steady current applied (mA/cm²)	Magnitude of V_t (in mV)	
	when $\dot{V}$ = min	when $\dot{V}$ = 80 V/sec
0.343	13.46	20.80
0.345	13.28	20.73
0.350	13.12	20.5
0.354	13.04	20.47
0.356	13.03	20.38
0.360	13.01	20.29
0.400	13.64	20.22
0.450	14.73	20.12
0.500	15.66	19.80
0.600	17.50	19.88
0.700	18.95	—
0.800	20.27	—

value at which, once it has been reached, the further action of the applied current is no longer necessary for spike generation. These workers considered that at that moment the inward and outward ionic currents become equal in absolute magnitude, regardless of whether a steady current or only a series of short pulses is used for stimulation. The calculations given below contradict this conclusion.

Changes in V, I_{Na}, I_l, and I_i, calculated for the case of stimulation of a nerve fiber by a pulse of steady current of threshold strength and 0.25 msec in duration, are shown graphically in Fig. 57a. The value of I_i reached its maximum 0.162 msec after application of the current, and it then immediately began to fall. At the end of the pulse I_{Na} was -0.305 mA/cm², while I_K and I_l were

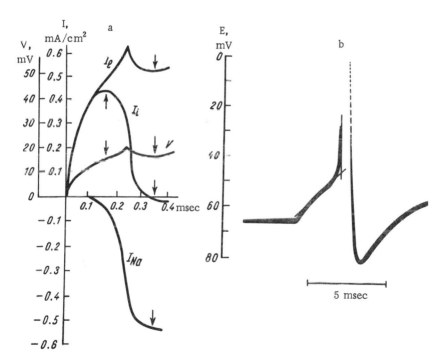

Fig. 57. Criterion of threshold depolarization of the membrane: a) relationships between ionic currents during stimulation of model of node by stimulus 0.25 msec in duration (I_m = 0.52 mA/cm²). Explanation in text; b) empirical determination of critical level of depolarization (after Baker, Hodgkin, and Meves, 1964).

0.005 mA/cm^2 and 0.672 mA/cm^2 respectively. The outward ionic current at this moment thus exceeded the inward I_{Na} by approximately 0.326 mA/cm^2. Nevertheless, discontinuing the stimulus ($I_m = 0$) did not prevent the action potential. Only the rate of increase of I_{Na} was slowed because of the smaller passive depolarization of the membrane.

There are considerable differences of opinion among investigators at the present time as regards the value which should be taken as the threshold potential.

During the application of short pulses of current Hodgkin, Huxley, and Katz (1949), Fatt and Katz (1951), Dodge (1961, 1963), Noble (1966), and Noble and Stein (1966) measured the threshold potential as the value of V at which $I_i = 0$. By contrast, Frankenhaeuser (1965) determined the threshold potential under the same conditions as the potential V at the end of the applied stimulus (V').

During the action of a steady current most workers judge the threshold potential either from the maximum value of the local response evoked by a near-threshold stimulus, or by the potential at which the depolarization of the membrane begins to rise steeply. This level is usually determined by eye or, as was done by Baker, Hodgkin, and Meves (1964) from intersection of lines drawn tangentially to the curves at the subthreshold and suprathreshold parts of the response (Fig. 57b). By contrast, Vallbo (1964) in experiments on single nodes of Ranvier, and Frankenhaeuser and Vallbo (1965) on a mathematical model of the membrane of the node, arbitrarily adopted the convention that the threshold potential is the value of V at which the rate of rise of regenerative depolarization reached 80 V/sec. The present writer followed a similar course in his own experiments (Khodorov, Bykhovskii, Korotkov, and Rusanov, 1966; Bykhovskii, Khodorov, Korotkov, and Rusanov, 1967). Finally, as already mentioned above, Noble and Stein (1966) considered that during the action of a steady current, just as during application of short stimuli, the threshold potential should be regarded as the potential at the time when I_i becomes equal to zero. All these differences of opinion can be explained by the fact that there is no uniformly accepted definition of the term "critical potential" in the current literature.

The dilemma from which a way out must be found is this: is the critical potential the level of depolarization of the membrane

which must be reached before a spike arises even if stimulation
has ceased, or must the critical potential be regarded as the level
of depolarization at which the local response grows into the action
potential?

In the first case, the value of the membrane potential E at the
end of action of the stimulus must be taken as the index E_c, for
after this value has been reached the subsequent course of events
inevitably leads to spike generation; in the second case it must be
the value of E at which the inward and outward ionic currents become equal in absolute magnitude (in response to short pulses of
current) or in their rates of increase (in response to a steady current).

The second interpretation of the terms "critical" and "threshold" potentials seems to the present writer to be more correct, for
the value of E_c at which the action potential begins directly reflects the active properties of the membrane and, in particular, the
state of the system regulating its sodium permeability.

An important argument in support of this view was put forward
by M. L. Bykhovskii: the nerve fiber begins to show properties of
negative resistance at the time when dI_i/dt reaches zero. From this
moment, an increase in V is no longer accompanied by an increase
in I_i, but the latter begins to "increase" negatively, i.e., it decreases and then changes its sign.

It is relatively simple to measure E_c or V_t experimentally if
changes in $\dot{V}$ are recorded simultaneously with changes in V (or E).

The suggestion of Frankenhaeuser and Vallbo (1965) that V_t
be measured as the value of V when the increase in $\dot{V}$ reaches 80
V/sec does not permit precise values of the threshold potential
to be determined. However, this method does yield sufficiently
reliable information about the character of the changes in V_t in
response to the action of various factors on the nerve fiber. Its
great advantage is that the value of V when $\dot{V}$ = 80 V/sec is approximately the same for different strengths of the stimulating
current (Table 2). The investigator thus has no need to choose a
current of very strictly rheobase strength in order to measure the
threshold potential.

In conclusion, the relationship between the threshold potential
and duration of the stimulus must be examined. The view is widely

held in the literature that the threshold (or critical) potential is independent of the strength and duration of the pulse of current, and is determined entirely by the properties of the membrane itself (Hagiwara and Oomura, 1958; Khodorov, 1965b; and others).

The author's calculations showed that certain substantial corrections must be introduced into this view. We have seen that in cases when the action potential arises after the end of the stimulus the threshold potential is in fact approximately the same despite differences in the duration of the stimulating pulses, provided that this duration is not too great and does not therefore give rise to a marked decrease in the value of h.

On the other hand, during the action of a steady current, the calculated value $V_t = 13.46$ mV is substantially below the value $V_t = 17.4$ mV measured by the short stimulus method. The reason for this difference is clear: if the action potential arises after the end of the stimulating pulse, the membrane can be depolarized only through the inward I_{Na}, and the spike cannot therefore begin until I_i has become equal to zero. In the case of stimulation by a steady current, on the other hand, critical depolarization of the membrane is brought about both by the ionic current and by the externally applied current, so that spike generation requires only equality between the rates of rise of the inward and outward ionic currents. This equality is reached, moreover, at significantly smaller values of V than equality between the values of the ionic currents.

Relationship between the Threshold Current and the Passive and Active Properties of the Membrane

The Rheobase

The Rheobase and Passive Properties of the Membrane. The threshold potential (V_t) of the membrane of a medullated nerve fiber is virtually independent of the leak conductance (g_l), whereas for a given value of the threshold potential, as a rule the higher the value of g_l the greater the rheobase strength of the current (Vallbo, 1964b; Frankenhaeuser and Vallbo, 1965).

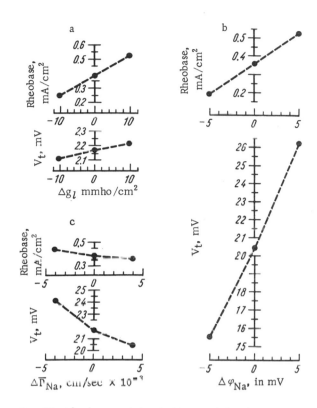

Fig. 58. Rheobase and threshold potential as functions of changes in the constants of ionic permeability. Abscissa:
a) change in g_l (Δg_l) compared with initial value (30 mmho/cm²); b) change in index of reactivity of sodium system $\Delta \varphi_{Na}$, in mV [made by changing the constants B in Eqs. (43) and (44)]; c) change in $\overline{P}_{Na}$ compared with its initial value of 8×10^{-3} cm/sec (Frankenhaeuser and Vallbo, 1965).

These relationships are seen particularly clearly when they are analyzed on a mathematical model of the nerve fiber membrane.

Changes in the threshold potential and rheobase during an increase and decrease of 10 mmho/cm² in the leak conductance, i.e., about one-third of its initial value, are shown in Fig. 58a.

The decrease clearly leads to a decrease of only 3% in the threshold potential, whereas the rheobase is lowered by 37%.

With an increase in g_l, the threshold potential and rheobase undergo the same change, but in the opposite direction.

The value of the rheobase is practically independent also of the capacitance of the excitable membrane. For instance, a two-fold increase in C_m from 2 to 4 μF/cm^2 increases the rheobase by only 6% over its initial value, while the threshold potential remains unchanged (Frankenhaeuser, 1965).

Relationship between Rheobase and Constants of Ionic Permeability of the Membrane. The effect of changes in the reactivity of the sodium activating system which, as mentioned above (see page 120), is determined by the rate constants α_m and β_m, on the rheobase is a matter of great interest.

It was shown in the previous chapter that V_t is a linear function of changes in the index of reactivity φ_{Na}.

As Fig. 58b shows, the same relationship also exists between the rheobase and φ_{Na}.

However, it will be noted that in this case the changes in rheobase are quantitatively greater than the changes in threshold potential. For instance, a decrease of 5 mV in φ_{Na} leads to a decrease of about 24% in V_t, whereas the rheobase falls by 44%. An increase of 5 mV in φ_{Na} increases the threshold potential V_t by 28% (from 20.4 to 26.15 mV), yet the increase in the rheobase under the same conditions is about 47% (Frankenhaeuser and Vallbo, 1964; Frankenhaeuser, 1965).

The evident explanation of this fact is as follows. It was shown in Chapter III that if a nerve fiber is stimulated by a steady current, critical depolarization of the membrane takes place as the result of summation of the passive electronic potential with the active (associated with an increase in P_{Na}) local response, the latter accounting for 30 percent or more of the total threshold change in potential. This fact was also reproduced in the model (Fig. 56b).

It will be perfectly clear that the greater the reactivity of the sodium transport system, the greater the contribution of the active component to critical depolarization.

Corresponding changes must also take place in the strength of the stimulating current required to produce critical depolariza-

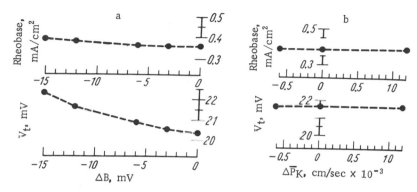

Fig. 59. The same as in Fig. 58. Abscissa: a) change in the constant B in Eq. (41). Values of h_∞ 0.825, 0.731, 0.599, and 0.4430 correspond to values of ΔB of 0, −5, −10, and −15 mV respectively; b) change in $\overline{P}_K$ compared with its initial value (0.54×10^{-3} cm/sec) (Frankenhaeuser and Vallbo, 1965).

tion of the membrane: with identical values of V_t, the greater the amount by which P_{Na} is increased for a given passive change in V, the lower the rheobase.

Let us now examine the relationships between the threshold potential and rheobase during a decrease or increase in the maximum sodium permeability $\overline{P}_{Na}$. It can be seen in Fig. 58a that an increase or decrease in $\overline{P}_{Na}$ is accompanied by approximately proportional changes in the threshold potential and threshold current. For instance, a decrease in $\overline{P}_{Na}$ by 5×10^{-3} cm/sec leads to an increase in the threshold potential and rheobase by about 11%; if $\overline{P}_{Na}$ is increased by the same amount, the threshold potential and rheobase are reduced by approximately 6%.

Similar relationships between the changes in V_t and I_t are also observed in the case of weakening of the inward sodium current through increased inactivation of the sodium system. For instance, with a decrease in h_∞ on account of the constant α_h, the threshold potential and rheobase are reduced by approximately the same number of times (Fig. 59a).

The maximum potassium permeability $\overline{P}_K$ has little effect on the values of the rheobase and threshold potential (Fig. 59b).

These changes are also very small in the case of a moderate increase in reactivity of the potassium system ($\Delta\varphi_K = -8$ mV) (Fig. 60). However, with a considerable decrease in φ_K the potas-

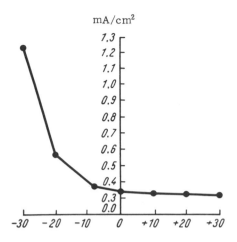

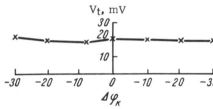

Fig. 60. Rheobase and threshold potential as functions of changes in reactivity of the potassium system. Abscissa, change in reactivity $\Delta\varphi$; ordinate: above, rheobase in mA/cm^2; below, threshold potential V_t in mV.

sium permeability of the resting membrane increases to such an extent that its net resistance falls, and for a threshold change in membrane potential to its previous value, a much stronger stimulating current has to be applied. This increase in rheobase takes place if φ_K is reduced by 20 mV and more (Fig. 60).

Threshold Current for Stimuli of Short Duration.

Dependence on Passive Properties of the Membrane

Capacitance of the Membrane (C_m). By contrast with the rheobase, the threshold current for short stimuli is closely dependent on the capacitance of the membrane. With an increase in C_m from 2 to 4 μF/cm^2 the threshold current for a stimulus with a current of 0.1225 msec, according to Frankenhaeuser's (1965) calculations, must increase by 41.5% (from 0.762 to 1.078 mA/cm^2), while the rheobase is virtually unchanged.

Such an increase in I_t for short stimuli is perfectly understandable: the greater the capacitance of the membrane for a given resistance, i.e., the higher the value of R_mC_m, the slower the potential will rise and, consequently, the stronger the current must be in order to charge the membrane during the time of its action up to the required level.

As a result, with an increase in $R_m C_m$ of the membrane, the entire strength versus duration curve is displaced toward high values of t. Conversely, with a decrease in $R_m C_m$, the thresholds for short pulses of current are lowered and the curve is shifted toward lower values of t.

This is the main explanation of the difference between the position of the strength versus duration curve relative to the time axis for excitable structures characterized by different RC values of the membrane.

For example, for nodes of Ranvier, for which $R_m C_m \sim 0.07$ msec, the constant $\tau_s = Q/I_0 = 0.15$ msec, whereas for squid giant axons with RC = 1.0 msec, τ_s is about 1.5 msec (at 20°C).

A similar difference is found between nerve fibers and the skeletal muscle fibers innervated by them, for whose membrane the value of $R_m C_m$ is approximately 20 msec (Nicholls, 1956; Maiskii, 1963).

The well known fact of the apparent lengthening of the chron-axie of a muscle after its curarization or denervation is connected with these same differences in the passive electrical properties of nerve and muscle fibers.

If a skeletal muscle, with its inervation intact, is stimulated by short pulses of current passed through extracellular elec-trodes, the nerve fibers, which have a shorter chronaxie (a smaller value of τ_s), respond first; the impression is thus created that the chronaxie of nerve and skeletal muscle fibers is the same (Lapi-que's "isochronism"). However, if neuromuscular transmission is blocked by curare, or after degeneration of the nerve fibers, the chronaxie is apparently considerably lengthened, because now the excitability of the muscle fibers themselves is tested. Experiments with stimulation via intracellular microelectrodes showed that, after application of curare in doses sufficient to block neuromus-cular transmission, the chronaxie of the muscle fibers is unchanged.

Resistance of the Membrane (R_m). A change in the resistance of the membrane has a twofold effect on the threshold density of the current for short stimuli.

First, an increase in R_m and, consequently, in $R_m C_m$ of the membrane slows the rise in V, which, other conditions being the same, would increase the threshold current. However, the increase in R_m leads at the same time to a marked increase in V = IR_m, for all durations of action of the current. Since this effect is dominant, the threshold current is reduced with an increase in R_m not only for long, but also for short stimuli. The influence of $R_m C_m$ is manifested only in the fact that a decrease in I_t is more marked at high values of t than at low.

The resistance of the node membrane at rest, as was mentioned earlier, is determined principally by the leak conductance. According to Frankenhaeuser's (1965) calculations a decrease in g_l from 30.3 to 15.5 mmho/cm^2 leads to a decrease of 52.9% in the rheobase, whereas the threshold current for a short stimulus (0.1225 msec) is reduced by only 36.9%. With an increase in g_l to 60.6 mmho/cm^2 the rheobase is increased by 118.7%, but the threshold current for the same short stimulus is increased by 78.0%.

Threshold Current for Short Stimuli as a Function of the Active Properties of the Membrane

Time Constant of Sodium Activation. Cooling a nerve fiber is known to cause a marked increase in the thresholds for short stimuli, and this effect is known to be connected with an increase in the utilization time (chronaxie). Since cooling the fiber within the range 20–5°C has no significant effect on the passive properties ($R_m C_m$) of the membrane, this change in the utilization time was naturally associated with the increase in duration of the time constant of sodium activation τ_m taking place under these circumstances.

The author's investigations conducted with A. M. Rusanov on the membrane of the node of Ranvier confirmed this hypothesis. The results showed that the time constant of sodium activation has a significant effect on the threshold strength of the current for short stimuli.

For instance, if τ_m is doubled [on account of a corresponding decrease in the values of the constants A in Eqs. (43) and (44)], the rheobase is increased by only 2.84% (from 0.342 to 0.352

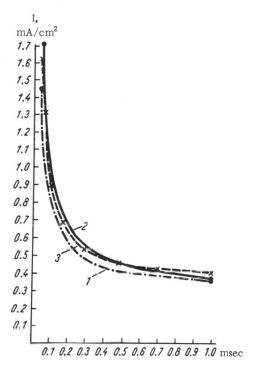

Fig 61. Theoretical strength versus duration curves for a node of Ranvier. Curves calculated: 1) for standard data of the model; 2) for a slowing of the sodium activation process (τ_m is doubled) by reducing the values of the constants A by 50% in Eqs. (43) and (44); 3) with an increase in the steady-state inactivation ($h = 0.443$) on account of a decrease in the constant B in Eq. (41) by 15 mV (Grilikhes, Rusanov, and Khodorov, 1967).

mA/cm^2), while I_t for stimuli of 0.01 and 0.001 msec in duration is increased by 18 and 30.2% respectively. The shift of the strength versus duration curve taking place under these circumstances is shown in Fig. 61.

An increase in the threshold strength of the current and, consequently, in the threshold quantity of electricity for very short stimuli, while the rheobase remained practically unchanged, led to an increase in the time constant of stimulation $\tau_s = Q/I_0$ from 0.15 to 0.19 msec, i.e., by 26.6%.

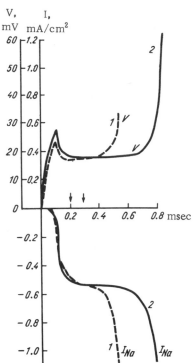

Fig. 62. Changes in membrane potential and inward sodium current in a single node of Ranvier calculated (1) for standard values of the constants of ionic permeability and (2) for slowing of activation of sodium permeability. Time constant τ_m is doubled. Duration of stimulus 0.1 msec. stimulating current in the first case is 0.816 mA/cm^2, in the second case 1.027 mA/cm^2. Arrows indicate time of reaching critical level of depolarization.

Analysis of the influence of τ_m on the kinetics of the ionic currents accompanying action potential generation reveals the cause of these changes.

The changes in V and I_{Na} calculated for stimulation of the node by threshold stimuli with a duration of 0.1 msec are shown in Fig. 62 for the initial value and for twice the initial value of τ_m. In the first case the threshold density of the current was 0.750 mA/cm^2, and in the second case 1.017 mA/cm^2.

The initial change in membrane potential (V') at the end of the stimulus for increased τ_m was clearly greater than when τ_m had the standard value (27.1 compared with 23.25 mV), while the critical level of depolarization (V$_t$) was virtually unchanged (17.5 mV).

The latent period of the action potential was considerably lengthened by an increase in τ_m. Under standard conditions of the model, equality of the inward and outward ionic currents was reached 0.2 msec after the end of the stimulus (first arrow), while after an increase in the value of τ_m equality was reached only after 0.3 msec (the second arrow).

The reason for these phenomena is clear. Since, in the case of short stimuli, the necessary increase in I_{Na} takes place only after cessation of the current, the slower the sodium permeability rises the higher must be the threshold strength of this current. The critical level of depolarization, however, is independent of the rate of increase of I_{Na}: only the time taken to reach it is changed, i.e., the latent period of the action potential.

Changes in the strength versus duration curve for a decrease in the strength of I_{Na} through inactivation of the sodium system are different in character.

Calculations showed that if inactivation is increased by the rate constant α_h, the thresholds for both long and short stimuli are increased by approximately the same factor, and the time constant of stimulation is thus unchanged.

As an example, results obtained with the node model for a change in h_α from 0.825 to 0.443 are given below. This decrease led to an increase in the threshold potential and the threshold strength of the current for all durations of stimuli by approximately 12%. Under these circumstances the strength versus duration curve was shifted parallel to itself along the voltage axis (Fig. 61). The time constant of stimulation τ_s, however, was virtually unchanged.

This result was a little unexpected, because an increase in the initial inactivation of the membrane (a decrease in h_∞) increases the rate of accommodation (see Chapter VII), and according to the accepted view (Hill, 1936), this ought to have a much greater effect on I_t for long stimuli than for short.

Maximal Sodium and Potassium Permeabilities. According to Frankenhaeuser (1965), a reduction by half in the standard value of $\overline{P}_{Na}$ (from 8×10^{-3} to 4×10^{-3} cm/sec) increases I_t for stimuli with durations of 1 msec and 0.1225 msec approximately equally. By contrast, doubling or reducing the value of the maximum potassium permeability $\overline{P}_K$ has no effect on the values of I_t for short and long stimuli.

Many investigations have been undertaken to analyze the effects of environmental agents (changes in ionic composition, the

action of local anesthetics and narcotics, depolarizing and hyperpolarizing currents, an increase or decrease in temperature, irradiation with infrared and ultraviolet light, and so on) on the
strength versus duration curve (Hill, 1936b; see the surveys by
Uflyand, 1941; Makarov, 1939, 1947; Golikov, 1950; Nasonov, 1959).

Unfortunately, the interpretation of the results of these investigations from the standpoint of modern views regarding the
nature of the utilization time is extremely difficult, if not impossible, because these workers did not take into account the influence
of the complex structure of excitable tissues on their results.

The role of this factor becomes particularly evident when the
action of an electric current on the nerve trunk is examined.

The nerve trunk consists of hundreds of thousands of nerve
fibers, differing in their excitability and surrounded by many connective-tissue layers and sheaths. Bishop (1928) and Grundfest
(1932) showed originally that these sheaths significantly disturb
the shape of the stimulus applied to the nerve trunk and, as a result, they modify the character of the strength versus duration
curve.

The following experiment, due to Tasaki (1939), is highly de-
monstrative in this respect. Tasaki measured the strength versus

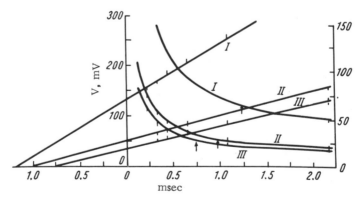

Fig. 63. Strength versus duration curves obtained for a frog nerve before (I) and after (II and III) removal of the connective-tissue sheath.
Arrows indicate chronaxie values; straight lines show quantity of electricity. Abscissa, time in msec; ordinate, voltage applied to nerve.
Temperature 8.7°C (Tasaki, 1939).

duration curve originally for an intact frog nerve trunk (Fig. 63, curve I), again after careful removal of the sheath at the site of stimulation (curve II), and finally, after careful removal of the connective-tissue septa separating the fibers (curve III).

At each stage changes thus took place in both the rheobase and the chronaxie.

This result could be explained in two ways: either by a decrease in the disturbance of shape of the stimulus applied to the nerve as a result of the decrease in quantity of connective tissue, or by changes taking place under these circumstances in the distribution of potential around the active nerve fiber. The second factor, as subsequent investigations by Tasaki (1939) showed, is the more important.

It can evidently be concluded from these results that precise measurements of the parameters of excitability are possible only in experiments on single excitable structures, provided always that the potentials are recorded directly from the points of stimulation of the fiber.

Chapter VII

Role of the Rate of Change of the Stimulus. The Phenomenon of Accommodation

The Accommodation Curve

The dependence of the effects of stimulation on the rate of rise of the electric current was known to the very first investigators who studied the laws governing the action of an electric current on excitable tissues (Du Bois Reymond, 1848; Pflüger, 1859; Fick, 1863; Engelman, 1870; Werigo, 1888) It was also established that with a decrease in the rate of rise of the stimuli below a certain limit no excitation whatever takes place.

The quantitative relationships between the duration of growth of the current and the threshold in frog nerve fibers were first studied by Lucas (1907), who used linearly increasing currents with different slopes. The lowest slope of the current at which propagated excitation still appeared was called by Lucas the "minimal gradient." This is also called the "critical slope."

The term "accommodation" was introduced into physiology by Nernst (1908), who considered that the increase in thresholds to a slowly growing current is due to chemical reactions taking place in living tissue during the prolonged passage of an electric current through it. Nernst based his theory on the results obtained by Kries (1884), who found sharp changes in accommodation under the influence of changes in temperature.

Accommodation of excitable tissues began to be studied intensively in the 1930's, after Hill (1936a) had put forward his mathematical theory of the "two time factors" in stimulation and methods for the precise measurement of parameters of currents growing as linear and exponential functions of time in physiological experiments became available.

Numerous experiments on cold-blooded and warm-blooded animals and on man showed that the accommodation curve characterizing the relationship between the threshold strength of current and the duration of its growth for different nerves, or for the same nerve in different states, can vary within wide limits. In some cases it approximates to a straight line, while in others the curve has an initial inflection and becomes a straight line only for long durations of stimulus growth; in a third group of cases it is initially a straight line, but then turns and becomes parallel to the time axis, exhibiting a "breakdown of accommodation."

The slope of the straight-line part of these curves relative to the time axis was usually used by investigators as a measure of the "rate of accommodation" or as its reciprocal — the time constant of accommodation (Solandt, 1936; Liesse, 1938; Scoglund, 1942; Granit and Scoglund, 1943; Kugelberg, 1944; Latmanizova, 1947, 1959; Khodorov, 1950a,b). The reasons for the differences in the course of the accommodation curve for different tissues were for a long time the subject of lively discussion between the investigators cited. However, none of them had any doubt that the large increase in thresholds observed in these experiments at the lower values of the slope of the current reflects true accommodation changes in the excitability of nerve fibers. However, when Tasaki and Sakaguchi (1950) investigated the accommodation of isolated nerve fibers of the Japanese toad, they found no increase in the thresholds with a decrease in the slope of a linearly increasing current. The threshold always remained approximately constant until the slope of the stimuli fell to a certain minimum (the minimal gradient), below which no action potential could be evoked, whatever the ultimate strength of this current.

Similar results were obtained on nodes of Ranvier of European frogs by Frankenhaeuser (1952) and on squid giant axons by Hagiwara and Oomura (1958).

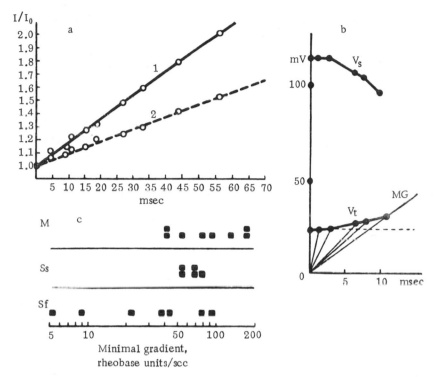

Fig. 64. Accommodation of a nerve and nerve fiber: a) accommodation curve of a frog nerve before (1) and after (2) removal of the connective-tissue sheath (Diecke, 1954). Abscissa, time of growth of current in msec; ordinate, threshold strength of current (I) in rheobase units (I_0); b) increase in threshold potential (V_t) and decrease in amplitude of action potentials (V_s) of a single node of Ranvier during a slow increase of the stimulating current. A gradient of linearly increasing current is shown by the slope of the thin lines drawn through the origin. The initial level of V_t is shown by the broken line. MG, minimal gradient, Abscissa, duration of growth of current; ordinate, values of V and V_t in mV (graph plotted from results obtained by Vallbo, 1964b; see Fig. 71a); c) minimum gradient of single nerve fibers of Xenopus laevis innervating muscle fibers (M), slowly adapting mechanoreceptors (Ss), and cutaneous mechanoreceptors (Sf). Numbers show values of minimal gradient in rheobase units/sec. Each square corresponds to one fiber (Vallbo, 1964b).

The impression was originally gained from the results of all these investigations that the accommodative increase in threshold is an artefact of the nerve trunk due to polarization of its connective-tissue sheaths (Sato, 1951; Diecke, 1954) (Fig. 64a) and to the recruiting of nerve fibers with different thresholds of stimulation and accommodation into the response to stimuli of different gradient (Sato, 1951). Nerve fibers themselves were characterized by the presence of a minimal gradient tens of times smaller in magnitude than the minimal gradient measured on the whole nerve (Lucas, 1907; Hill, 1936a).

However, the subsequent more detailed investigations of Diecke (1954) and Vallbo (1964b) on single nodes of Ranvier cleared up a number of important points concerning these views. They showed that an accommodative increase in threshold also takes place in isolated fibers, but it is much weaker than in the whole nerve. To detect this increase in threshold, stimuli with slopes very close to the minimal gradient have to be used. In this case, an increase of threshold by 20–25% over the rheobase can be found (Fig. 64b).

Proof that during slow increase of a depolarizing current the thresholds of stimulation of a nerve fiber rise steadily can also be obtained in another way.

The experiment is carried out as follows. The minimal gradient (critical slope) of the test fiber is first determined. A linearly increasing current with a gradient very slightly lower than this minimal gradient is then applied, and against its background, the thresholds are measured by means of test stimuli of short duration applied at different intervals after application of the current.

In these experiments, approximately 15 msec after application of the linearly increasing current the thresholds were seen to increase gradually, and to reach (Diecke, 1954) approximately 175% of their initial value.

These results show that polarization of the membranes and recruiting of different fibers into the response in experiments on the whole nerve trunk influence only the quantitative aspect of the accommodative growth of thresholds, and do not cause this phenomenon as investigators had first considered.

In response to stimulation of single nerve fibers by linearly increasing currents of small gradient, besides a small increase in the thresholds, a marked decrease in the amplitude and rate of rise of the action potential also was found (Frankenhaeuser, 1952; Diecke, 1954; Vallbo, 1964b) (Figs. 64b and 71a).

The minimal gradient of various motor and sensory nerve fibers of amphibians was measured experimentally by Vallbo (1964a) and by Bergman and Stämpfli (1966).

In Vallbo's (1964a) experiments, conducted on the African toad Xenopus laevis, the minimal gradient of different nodes of Ranvier varied from 5 to 200 rheobase units/sec. It was much higher (113 ± 43 rheobase units/sec) than in sensory fibers, whether connected with fast-adapting cutaneous mechanoreceptors (66 ± 43 rheobase units/sec) or with slow-adapting muscle mechanoreceptors (85.2 ± 23.3 rheobase units/sec (Fig. 64c).

Bergman and Stämpfli (1966) carried out experiments on nodes of Ranvier of nerve fibers of European frogs (Rana esculenta) and also found very considerable differences between the minimal gradients of motor and sensory fibers: 244 ± 14 rheobase units/sec for motor fibers and 89 ± 13 rheobase units/sec for sensory fibers. They were unable to correlate these differences with the state of the potassium permeability system of the membrane of these fibers.

In motor fibers the reactivity of the potassium system was found to be much higher than in sensory fibers.* For this reason, when tetraethylammonium is applied to nodes of Ranvier in motor fibers there is a much greater decrease in the minimal gradient (by 5.6 ± 0.2 times) than in sensory fibers (by 2.5 ± 0.5 times). As a result, after treatment with tetraethylammonium, the minimal gradients of the two sets of fibers become very similar: the minimal gradient in motor fibers becomes close to 47 rheobase units/sec, compared with 37 rheobase units/sec for sensory fibers (Fig. 65) (see also Bergman, 1969).

No systematic data for accommodation of single peripheral nerve fibers of warm-blooded animals can yet be found in the recent literature. Investigations of accommodation conducted on thin bundles of fibers, however, cannot be used to describe this phenomenon quantitatively. Nevertheless, these results are inter-

---

*During depolarization of the membrane of a node of Ranvier by 20 mV, P_K in the motor fibers is approximately twice as high as in sensory fibers, although the maximal values of P_K ($\overline{P}_K$) in both are practically identical.

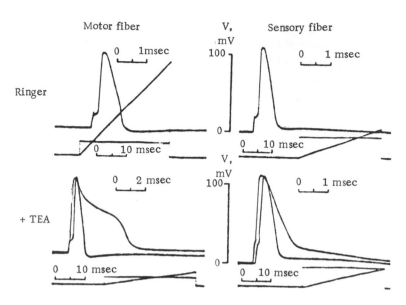

Fig. 65. Action potential and critical slope in motor and sensory fibers. Top row: fiber in ordinary Ringer's solution. Bottom row: fiber in Ringer's solution containing 5 mM TEA. Note differences in course of potential in motor and sensory fibers and very powerful effect of TEA on spike duration and critical slope in the motor fiber. Recording from single node of Ranvier by double air-gap method (Bergman, 1969).

esting for the comparative analysis of accommodation of motor and sensory fibers.

Scoglund and Granit (Scoglund, 1942; Granit and Scoglund, 1943) showed that the rate of accommodation (measured from the slope of the accommodation curve) for motor nerve fibers is much greater than the rate of accommodation of sensory fibers. This correlates closely with the more marked ability of sensory fibers to give prolonged discharges of spikes in response to stimulation by a steady current (see page 201).

In experiments on nerves of warm-blood animals and man, the above-mentioned phenomenon of a breakdown of accommodation was found in response to the action of slowly growing stimuli.

In its outward manifestation this phenomenon is diametrically opposite to the minimal gradient phenomenon. The essence of this phenomenon is that the accommodative increase in threshold ceases as the rate of increase of the current is slowed as soon as its gradient falls below a certain minimum. In motor nerve fibers of warm-blooded animals, this curve forms a plateau at the level of 3-4 rheobase units; in sensory fibers it does so at 1.5 rheobase units (Granit and Scoglund, 1943).

A "breakdown of accommodation" has also been found for human nerves during stimulation by linearly increasing currents through electrodes applied to the skin. For instance, Kugelberg (1944) found that the accommodation curve reaches a plateau at 2.5-5 rheobase units in the case of motor fibers and 1.5-2 rheobase units in the case of sensory fibers.

If the strength of the stimulating current is varied widely enough, a breakdown of accommodation appears in certain cases in frog nerves also (Khodorov, 1950b).

Since this phenomenon was never observed in single nerve fibers, the idea naturally arose that it is due to recruiting of nerve fibers with a high threshold and with a virtually zero minimal gradient in the response to slowly increasing stimuli (Sato, 1951).

Accommodation of the bodies of nerve cells and of their axons has been investigated by the use of intracellular microelectrode techniques. However, during stimulation through the microelectrode, there is a very complex distribution of potential between regions of the neuron membrane differing in their passive and active properties (Eccles, 1959; Rall, 1964). Analysis of results obtained by these techniques is impossible without taking into consideration the cable properties of the test object

The problem of accommodation of the neuron body is therefore best examined in the next chapter, in which we deal with the influence of the microgeometry of excitable structures on parameters of excitability and of the action potential.

However, it may be mentioned here that stimulation of intracerebral nerve fibers through intracellular (Sasaki and Otani, 1961; Bradley and Somjen, 1961) or extracellular (Ushiama, Koizumi, and Brooks, 1966) microelectrodes gave results which are similar in principle to those obtained previously on isolated nerve fibers.

The existence of a minimal gradient and the relative constancy of the threshold of stimulation have been demonstrated for stimuli with a slope exceeding the minimal gradient. A slight increase in the threshold was discovered only close to the minimal gradient.

The value of the minimal gradient for motor nerve fibers of the anterior columns of the cat spinal cord, according to Ushiama, Koizumi, and Brooks (1966), is about 200 rheobase units/sec. Posterior column fibers have a smaller minimal gradient, from 15 to 100 rheobase units/sec.

These results were obtained by stimulation of nerve fibers and recording potentials from them by means of extracellular microelectrodes introduced beneath the myelin sheath (Tasaki, 1952). With the microelectrode tip in this position, the fiber was stimulated by an anodic current, and under these circumstances excitation arose in the next node under the influence of the outward current.

In the experiments of Sasaki and Otani (1961) the minimal gradient of axons inside the spinal cord varied from 18.4 to 57.7 rheobase units/sec (Fig. 65b).

However, all investigators have observed that insertion of a microelectrode inside the axon causes injury to it, so that these values of the minimal gradient are possibly a little too high.

The Mechanism of Accommodation

Werigo (1888) first postulated that the accommodation phenomena are based on "early cathodic depression," developing in the nerve while the current grows to the threshold level (Chapter I). Werigo illustrated his view graphically as follows.

In Fig. 66a, line ACC' corresponds to a square pulse of current, line AE to a linearly increasing current, and line AB to the height of the threshold before the beginning of the action potential. On application of an instantaneously growing current of above-threshold strength, excitation takes place only during the first few moments of its action, for immediately after "make" the excitability of the nerve under the cathode begins to decline (the increase in threshold is shown by the line BR), as a result of which the current applied becomes subthreshold in magnitude.

If the current grows slowly the threshold does not remain unchanged, but also rises steadily (the line BF), so that at the time

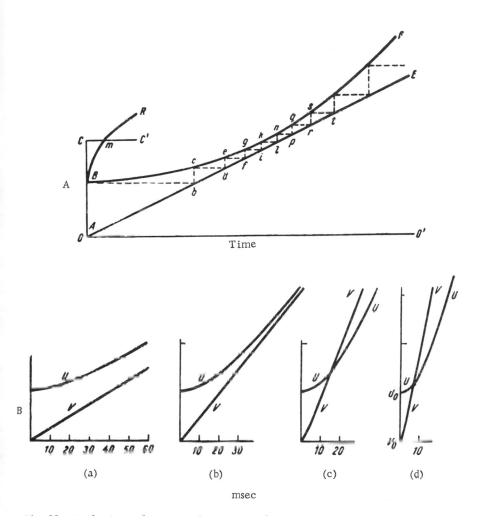

Fig. 66. Mechanisms of accommodation: A) after B.F. Werigo (1888). ACC', square pulse of current; AE, linearly increasing current; AB, height of threshold of nerve stimulation before application of linearly increasing current; bc, ed, gf, ki, nl, gp,..., thresholds for testing stimuli during period of increase of current AE; B) mechanism of accommodation of nerve to linearly increasing current of different gradient (a-d) after Hill (1936a). Excitation arises at the time when the potential V reaches the threshold level U. a-d) Changes in I with different rates of rise of V.

when the linear current reaches the former threshold strength at the point b the threshold is now higher than this level by the amount bc, and so on. In the case represented on the graph, the linearly increasing current cannot catch up with the ever-increasing threshold, and it cannot therefore evoke excitation.

A similar idea of the mechanism of accommodation was developed 50 years later by Hill (1936a), who put forward a mathematical theory of stimulation and accommodation. However, since Hill gave a purely formal description of the changes in potential and threshold during the action of stimuli of different duration and gradient on a nerve (Fig. 66b), this theory is now only of historical interest.

An explanation of the ionic mechanisms of accommodation became possible with the advent of the modern membrane theory of the nervous impulse.

Action potentials of a node of Ranvier, calculated by the equations of Frankenhaeuser and Huxley (1964) for different initial values of the membrane potential, are given in Fig. 67a.*

Depolarization of the membrane leads to an increase in the critical potential and a decrease in the amplitude, rate of rise, and duration of the action potential. The stronger the depolarization of the membrane (the higher the initial value of V), the more marked these changes become. With an increase in the potential inside the membrane by $V = 20$ mV ($E = -50$ mV), the calculated responses become graded in character (see curves 3 and 4 in Fig. 67a).

The relationships between the critical potential, amplitude of the spike, and the maximal rate of rise, on the one hand, and the initial level of the membrane potential, on the other hand, established in these investigations are shown graphically in Fig. 67c. They are very similar to those observed under analogous experimental conditions in the single node of Ranvier (Fig. 3, curve 1).

At the same time, the calculations give some idea of the changes taking place under these conditions in P_{Na}, P_K, I_{Na}, I_K, and the inactivation constant h during generation of action potentials (Fig. 68).

---

*These calculations were made jointly by the author and A. M. Rusanov.

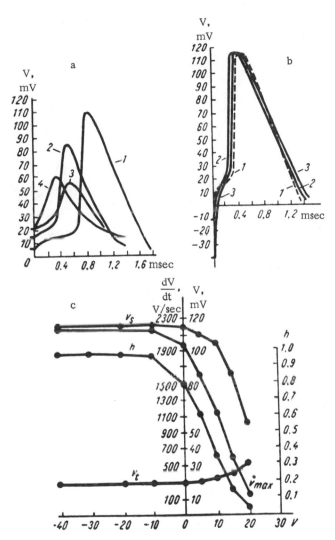

Fig. 67. Calculated action potentials of a node of Ranvier for different initial levels of the membrane potential: a) effect of depolarization. Stimulating current I_m = 0.5 mA/cm^2; b) effect of hyperpolarization; I_m = 0.7 mA/cm^2; c) amplitude of spike (V_s), maximum rate of its rise ($\dot{V}_{max}$), threshold potential (V_t), and inactivation constant (h) as functions of the change of potential toward hyperpolarization (an increase in V) or depolarization (a decrease in V).

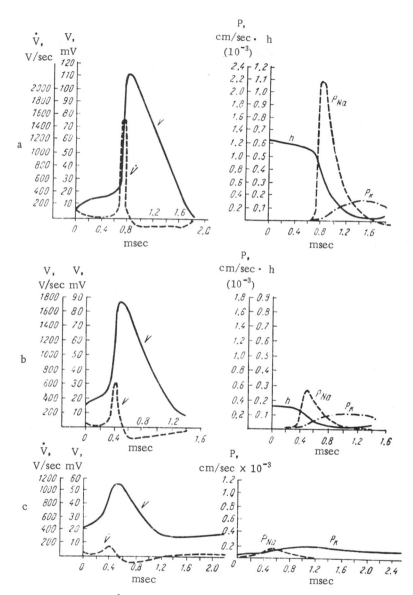

Fig. 68. Changes in V, V̇, P_{Na}, P_K, and h during the development of an action potential, calculated for different initial levels of the membrane potential. Shifts of resting potential toward depolarization: 5 mV (a), 15 mV (b), 20 mV (c); in case c, h_∞ was 0.056. Model of membrane of node of Ranvier.

Prolonged depolarization of a membrane, according to Hodgkin's theory, gives rise to two types of changes in the ionic permeability of the membrane: inactivation of the sodium channels and an increase in permeability for K^+ ions.

It can be concluded from the following facts and calculations that the cathodal increase in the critical potential and decrease in the amplitude of the action potential in the node of Ranvier are due mainly to inactivation of the sodium transport system, and that an increase in potassium permeability has no significant effect on these changes in the electrical activity of the node of Ranvier.

1. Inhibition of activation of the potassium permeability of the membrane by tetraethylammonium does not weaken the decrease in amplitude of the action potential and the increase in the critical level of depolarization observed under the cathode, as would take place if the increase in P_K made a significant contribution to the development of cathodal depression (Khodorov, 1965; Khodorov and Belyaev, 1965a).

2. All procedures reducing initial sodium inactivation (an excess of Ca^{++} ions in the medium, addition of Ni^{++}, Co^{++}, or Cd^{++} to the solution, strong preliminary hyperpolarization of the membrane of the node, etc.), weaken the catelectrotonic changes in the critical level of depolarization and action potential of the nerve fiber (Frankenhaeuser, 1957b, Takahashi, Usuda, and Ehara, 1962; Khodorov and Belyaev, 1963b; Khodorov, 1965a).

3. Even a large increase in P_K (by changing the rate constants α_n and β_n) has virtually no effect on the calculated value of the critical potential of the node membrane. The amplitude of the action potential is only slightly reduced under these circumstances (see Fig. 16b).

4. Changes in the calculated values of the critical potential and in the amplitude of the action potential of the model node during depolarization are almost completely identical with those resulting from a corresponding increase in the initial inactivation (a decrease in h_∞), while the values of the resting potential and the constants of ionic permeability of the membrane remain unchanged.

The last of these points requires some explanation.

   Depolarization of the membrane increases inactivation of the sodium system (lowers h) as a result of an increase in the rate constant β_h and a decrease in the rate constant α_h (see Fig. 28a). However, the same change in these constants can also take place without changes in the membrane potential, through a decrease in the values of the constants B by the appropriate number of milli- volts in Eqs. (41) and (42). In this case the potassium permea- bility of the membrane remains at its previous level, different from what takes place during depolarization of the membrane.

   In Fig. 69 the values of the critical potential, the amplitude of the action potential, and the maximal rate of rise are plotted in a system of coordinates against the corresponding values of h_∞ obtained by a change in membrane potential (continuous lines) and of changes in the constants α_h and β_h (broken lines). It is clear that the curves coincide almost completely.

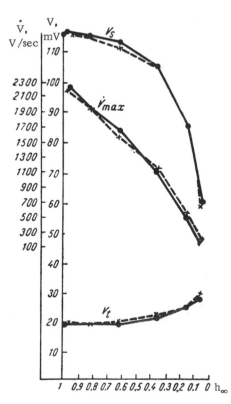

Fig. 69. Amplitude of action potential (V_s), maximal rate of rise ($\dot{V}_{max}$), and threshold potential (V_t) as functions of the initial level of inactivation of sodium permeabil- ity. Continuous lines represent the situa- tion when the level of inactivation is changed through a change in the initial level of V; broken lines when the change takes place through a change in the values of the constants B by the same number of millivolts in Eqs. (41) and (42). Abscissa, value of h_∞; ordinate, V_s, $\dot{V}_{max}$, and V_t.

An increase in potassium permeability of the membrane during depolarization affects mainly the threshold strength of the current and the duration of the descending phase of the action potential, so that the threshold current rises much more strongly during catelectrotonus than the threshold potential.

The hypothesis that accommodation of excitable structures to a slowly growing current is based on the same mechanism as cathodic depression (Werigo, 1888) was originally tested in experiments on whole frog nerves (Khodorov, 1950a,b).

These experiments showed that under the influence of a wide variety of treatments (with Ca^{++} and K^+ ions, with monoiodoacetate and eserine, cooling and warming) the rate of accommodation of the nerve to an exponentially increasing current, measured by Solandt's (1936) method, and also the rate of increase in the threshold under the cathode of a subthreshold steady current undergo parallel changes.

This result was interpreted as evidence in support of the view that cathodic depression and accommodation are identical in nature. Subsequent investigations on single nerve fibers fully confirm this conclusion.

The results obtained by Vallbo (1964b) when measuring accommodation of single nodes of Ranvier from frog nerve fibers have already been given above (see page 181).

The special feature distinguishing Vallbo's experiments was that measurement of accommodation was combined on the same node of Ranvier with analysis of the ionic currents undertaken by clamping the membrane voltage. As the index of inactivation, Vallbo used the parameter V_h, which shows by how many millivolts the initial resting potential of the fiber must be lowered in order that the sodium system be inactivated by half, i.e., the value of h reduced to 0.5 (see page 87).

In Vallbo's experiments, V_h varied from 2.8 to 28 mV, and the threshold potential V_t from 23.2 to 33 mV. Approximately linear relationships were established between the logarithm of the critical gradient and the difference $V_h - V_t$ (Fig. 70).

Vallbo (1964b) concluded from his investigations that these differences between the degree of initial inactivation $(1 - h_\infty)$ of

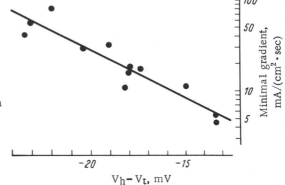

Fig. 70. Minimal gradient as a
function of $V_h - V_t$. Explanation
in text (Vallbo, 1964).

the frog nerve fibers are due mainly to variation of the rate-
constant α_h; the rate constant β_h varies comparatively little among
different fibers (see Fig. 28a).

Vallbo found no correlation between the accommodation of
nerve fibers and the constants of sodium and potassium permea-
bility $\overline{P}_{Na}$ and $\overline{P}_K$, or between the accommodation and the leak conductance g_l. He therefore postulated that the differences in the
values of the rate constant α_h are the main factor determining
normal variations in the minimal gradient for different nerve
fibers of the frog.

To test this hypothesis and to make a more detailed quantitative study of the relationship between accommodation of the nerve
fiber and the various constants of ionic permeability of the membrane, Frankenhaeuser and Vallbo (1965) made use of a mathematical model of the membrane of the node of Ranvier.

Their investigations showed that this model reproduces not
only the effects of stimulation of a nerve fiber with square pulses
satisfactorily, but also reproduces the basic principles of accommodation. During the action of linearly increasing stimuli the
model exhibited a well-marked minimal gradient which, at $h_\infty =$
0.7073, was very close in magnitude to that found in experiments
on an actual node of Ranvier.

At slopes slightly above the minimal gradient, action potentials were generated. These were characterized by reduced amplitude and by increased threshold potential, i.e., by changes

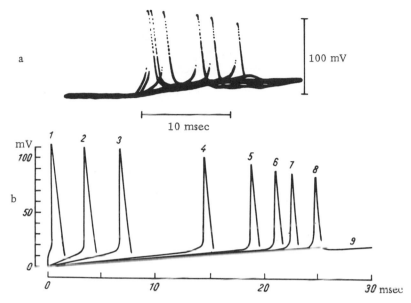

Fig. 71. Experimentally observed (a) and calculated (b) changes in membrane potential in response to a square pulse of current (1) and to linearly increasing stimuli of different gradient (2-9). All data for the calculations were standard except the constant B in Eq. (41), which was reduced from −10 mV to −16 mV (h = 0.707) (Frankenhaeuser and Vallbo, 1965).

which also were similar to those observed experimentally (Fig. 71). The slight quantitative differences between the experimental and calculated values did not exceed those expected, having regard to the well-known inaccuracies of the method of analysis of ionic currents in a membrane under voltage-clamp conditions.

Investigations on a nerve model also enabled the kinetics of of ionic permeabilities and currents passing through the membrane of the node of Ranvier to be analyzed during the action of linearly increasing stimuli with different gradients. The results showed clearly that the accommodative increase in threshold potential and decrease in the amplitude and rate of rise of the action potential are due to inactivation of the sodium permeability (a decrease in h_∞) developing during the ascending phase of the stimulus. For this reason, the lower the initial level of inactivation and the faster it develops during depolarization of the membrane, the more marked the accommodation.

However, this investigation did not fully elucidate the nature of the concrete mechanism of total disappearance of the action potentials in the nerve fiber with a decrease in the rate of rise of the current below the minimal gradient.

There were two possible, but slightly different, explanations of this phenomenon, called "Einschleichen" (stealing in) by German authors.

According to one explanation, during the slow rise of the current the mechanism of action potential generation is totally inactivated, so that when V has risen to the nominal threshold value the fiber is already completely unable to generate spikes. This hypothesis was put forward by Frankenhaeuser as long ago as in 1952.

The other explanation is that under the influence of inactivation of P_{Na} and an increase in P_K, there is a concomitant increase in the threshold potential, with the result that the membrane depolarization induced by the slowly rising current remains subthreshold all the time. This interpretation of the mechanism of accommodation is similar, in principle, to that given by Werigo (1888), Nernst (1908), Hill (1936a), and others.

To examine which of these explanations corresponds closer to reality the following investigations were carried out on a mathematical model of the membrane of the node of Ranvier (Bykhovskii, Korotkov, Rusanov, and Khodorov, 1966).

Against the background of a linearly increasing current of subminimal gradient, at various time intervals (10, 20, 25, 31, and 40 msec) after its beginning, short testing stimuli were applied to the "membrane." By means of these stimuli, the thresholds and the character of the response of the model to stimulation were determined (Fig. 72a).

It will be clear that during gradually increasing depolarization the critical potential E_c rises and the amplitude of the action potential definitely falls. Under these circumstances the model remains able to generate an action potential in response to the testing stimulus not only at the 25th millisecond of action of the current, i.e., when a current of subminimal gradient evokes only a local response, but also at the 31st millisecond of depolarization, despite the fact that by this time h has fallen from 0.825 to 0.038. Admitedly, responses at such a high level of inactivation of the sodium system have become graded in character (Fig. 72b).

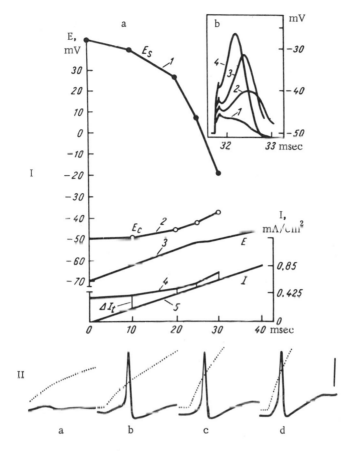

Fig. 72. Mechanism of the "stealing in" phenomenon. I: a) Change
in the peak value of the action potential (1), critical potential (2),
and resting potential (3) during a linearly increasing current (5) of
subminimal gradient (21.24 mA /cm²/sec). Testing stimuli, 1 msec
in duration, were applied 10, 20, 25, 31, and 40 msec after the beginning of the linearly increasing current. I_t, threshold strength of
testing stimulus; 4) summary density of linearly increasing and testing currents. Asterisk denotes values of current and potential at
which responses of the model became gradual in character. In the
inset (b) the gradual responses obtained by application of short
(0.1225 msec) stimuli of continually increasing density [1) 0.2
mA /cm²; 2) 0.25 mA /cm²; 3) 0.3 mA /cm²; 4) 0.4 mA /cm²] are
shown (Bykhovskii, Korotkov, Rusanov, and Khodorov, 1966); II:
changes in membrane potential of the squid giant axon when stimulated by linearly increasing stimuli of different gradients (a-d)
by means of an axial wire electrode inserted into the fiber. Calibration: voltage 50 mV, time 2 msec. Explanation in text (Hagiwara and Oomura, 1958).

Precise determination of the threshold is impossible when graded action potentials appear. However, judging from the action potential obtained in response to stimulation at 0.3 mA/cm^2, the critical potential E_c at the 31st millisecond of increase of the current was about -42 mV, i.e., 8 mV higher than the initial level of E_c (about -50 mV).

The system completely lost its ability to generate responses to additional stimuli only at the 40th millisecond of increase of the current, when h had fallen to 0.011.

The results obtained with the model thus indicate that the main cause of the accommodation phenomenon in the node of Ranvier is an increase in the critical potential, preceding the slowly increasing depolarization of the membrane. This conclusion is in agreement with the results of Diecke's (1954) experiments.

Calculations also showed that this increase in the critical potential is almost entirely due to inactivation of the sodium system. The role of an increase in P_K in this process is very small (see page 189). The increase in P_K during growth of the stimulating current mainly affects the threshold strength of this current, for during an increase in P_K the net resistance of the membrane falls and, consequently, the rate of passive depolarization of the membrane decreases (see Fig. 72, curve 3).

The effect of changes in P_K on the minimal gradient is seen particularly clearly when accommodation is investigated in the squid axon, in which the potassium conductance of the active membrane is much greater than the leak conductance (see page 136). The records obtained by Hagiwara and Oomura (1958) for changes in the membrane potential of the squid giant axon during its stimulation by linearly increasing currents of different gradients are given in Fig. 72b (in these experiments the nerve was stimulated through an axial intracellular electrode). In cases in which the rate of increase in strength of the current exceeded the minimal gradient the action potential appeared when depolarization of the membrane reached the critical level (Fig. 72, II,b,c,d).

During the action of a current of subminimal gradient (Fig. 72, II,a), however, the delayed rectification (associated with an increase in g_K) led to such a marked reduction of this depolarization that it was not strong enough to reach the critical level within the specified time.

In medullated nerve fibers the role of an increase in the potassium permeability of the membrane in the phenomena of accommodation was demonstrated by the experiments of Bergman and Stämpfli (1966), mentioned above, in which tetraethylammonium was applied to nodes of Ranvier. Inhibition of the growth of P_K, as we have seen above (see page 189), in these experiments led to a marked decrease of the minimal gradient.

Minimal Gradient as a Function of the Constants of Ionic Permeability of the Membrane

Frankenhaeuser and Vallbo (1965) calculated the changes in minimal gradient as a function of the various constants of ionic permeability of the excitable membrane.

In the first place, the relations discovered between accommodation and the initial steady-state level of inactivation of the sodium system, discovered in experiments on actual nodes, were confirmed.

An investigation on the model showed that normal variations in the minimal gradient of frog nerve fibers can in fact be but insfactorily explained by differences in the initial level of sodium inactivation.

The minimal gradient is shown in Fig. 73a as a function of h_∞, changes in which were produced by changes in the constant α_h. It is clear that a decrease in h_∞ below its initial value leads to a marked increase in the minimal gradient, i.e., to the more rapid development of accommodation.

With an increase in h_∞ (not shown in Fig. 73), on the other hand, the minimal gradient falls or even disappears completely: responses of the fiber to a very slowly increasing current assume the character of oscillations. Small oscillations begin when the membrane potential changes by about 17 mV. They gradually increase in amplitude and assume the form of small action potentials when the gradient of the current is higher.

To compare the results of the calculations with experimentally obtained values of the minimal gradient, these workers plotted both in a system of coordinates against the potential difference $V_h - V_t$.

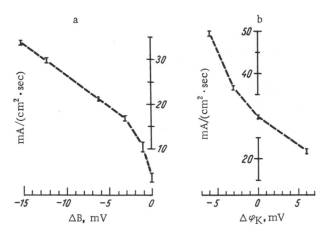

Fig. 73. Minimal gradient as a function of the initial level of
inactivation (a) and sensitivity of the potassium system of depolarization (b). a) Abscissa: changes in the constant B (ΔB)
in Eq. (41) in mV. Values of ΔB (0, −5, −10, −15 mV) correspond to the following values of h_∞: 0.825, 0.731, 0.599, and
0.443. Ordinate: minimal gradient in mA/cm^2/sec; b) abscissa: changes in index of reactivity of potassium system ($\Delta\varphi_K$)
in mV. A decrease in $\Delta\varphi_K$ corresponds to an increase in reactivity. Ordinate: minimal gradient in mA/cm^2/sec (Frankenhaeuser and Vallbo, 1965).

The similarity between the curves was obvious. The slight differences between them were attributed by Frankenhaeuser and Vallbo
to the fact that the curve of h_∞ versus V is steeper in the case of
the nerve model than for actual nerve fibers.

   Calculations also showed that the state of the potassium system may also have a substantial effect on the minimal gradient.
An increase in its reactivity (a decrease in φ_K, see page 132),
causes a marked increase in the minimal gradient (see Fig. 73b).
Variation of the maximal sodium and potassium permeabilities
$\overline{P}_{Na}$ and $\overline{P}_K$ within the range from 4 to 12 and from 0.6 × 10^{-3} to
2.4 × 10^{-3} cm/sec respectively leads to only moderate changes
in the minimal gradient. A decrease in $\overline{P}_{Na}$ or an increase in $\overline{P}_K$
increase the rate of accommodation, while a decrease in $\overline{P}_K$ or an
increase in $\overline{P}_{Na}$ cause accommodation to take place more slowly
(Fig. 74a,b).

   The relationship between the minimal gradient and reactivity
of the sodium system has not yet been studied.

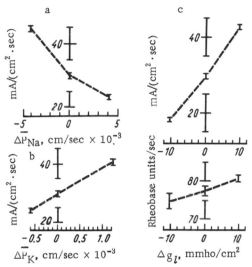

Fig. 74. Minimal gradient as a function of changes in maximal sodium permeability, maximal potassium permeability, and leak conductance. In a) abscissa, $\Delta \bar{P}_{Na}$ in cm $\cdot$ sec^{-1} $\times$ 10^{-3}; ordinate, minimal gradient in mA/cm^2/sec; b) abscissa, $\Delta \bar{P}_K$ in cm $\cdot$ sec^{-1} $\times$ 10^{-3}; ordinate, minimal gradient in mA/cm^2/sec; c) abscissa, Δg_l in mmho/cm^2; ordinate, minimal gradient: top in mA/cm^2/sec, bottom in rheobase units/sec. Model of membrane of node of Ranvier (Frankenhaeuser and Vallbo, 1965).

Changes in the leak conductance (g_l) have a moderate effect on the minimal gradient, measurable in units of mA/cm^2/sec; this effect is much weaker if the minimal gradient is measured in rheobase units/sec (Fig. 74c).

This will become clear if it is remembered that the leak conductance, which in fact determines the resistance of the resting node of Ranvier, has only a very slight effect on the threshold potential, but a strong effect on the threshold strength of the current (Fig. 60).

These calculations thus showed that nearly all the constants of ionic permeability of the membrane have some influence on the minimal gradient. Changes which reduce the outward ionic current or increase the inward ionic current at near-threshold

potentials lead to a decrease in the rate of accommodation -- a decrease in the minimal gradient. Conversely, changes which
strengthen the outward or weaken the inward ionic currents increase the maximal gradient, i.e., increase the rate of accommodation.

These predictions of the theory are important to the analysis
of the experimental data now available pertaining to the influence
of different environmental factors on the rate of accommodation.

The rate of accommodation is known to increase in catelectrotonus and to decrease in anelectrotonus. These effects were obtained initially in experiments on whole nerve trunks (Fabre, 1936;
Davidov, 1943; Zhukov, 1940), and later on isolated nerve fibers
(Tasaki, 1950).

From the standpoint of the Hodgkin—Huxley theory, the cause
of the electrotonic changes in the rate of accommodation is that
prolonged depolarization of the membrane strengthens inactivation
of the sodium system (fall of h_∞) and increases potassium permeability. Hyperpolarization of the membrane, on the other hand,
weakens or even abolishes the inactivation and lowers K^+ permeability.

An increase in the concentration of potassium ions in the medium produces the same effect on accommodation as a cathodal
current (Solandt, 1936; Magnitskii, 1938; Khodorov, 1950a,b). By
depolarizing the membrane, and thus strengthening inactivation
of the sodium system, K^+ ions considerably increase the rate of
accommodation of nerve fibers.

Weakening or strengthening of the inward sodium current can
be produced not only by a change in the level of inactivation of the
sodium system, but also by a decrease or increase in the Na^+
ion concentration in the medium.

In accordance with predictions of the theory, Diecke's (1954)
experiments on single medullated fibers showed that a decrease
in Na^+ ion concentration in the medium from 114 to 27.8 mM leads
to an increase in the minimal gradient (by about 150%), while an
increase in the Na^+ ion concentration, on the contrary, reduces the
minimal gradient (by about 60%).

If accommodation is in fact connected with the inactivation of
P_{Na} and the delayed increase in P_K, a decrease in the rate of de-

velopment of these processes must reduce the minimal gradient, and vice versa.

This hypothesis is confirmed experimentally: cooling a nerve fiber considerably increases the time constants of sodium inactivation τ_h and potassium activation τ_n ($Q_{10} \sim 3$). At the same time, the minimal gradient is reduced on cooling. On heating, the opposite change takes place: τ_h and τ_n are reduced and accommodation takes place more rapidly (Tasaki, 1949; Khodorov, 1950a,b).

Accommodation and Repetitive Responses

The study of the mechanism of accommodation has shed light on the nature of repetitive responses.

Most nerve cells of invertebrates and vertebrates have the ability to respond by regular discharges of spikes to the action of a steady depolarizing current. Investigation of these repetitive responses is of great interest, because they are basically similar in nature to the repetitive discharges arising in a neuron under natural conditions during prolonged depolarization of the membrane by a synaptic current (in the case of prolonged excitatory, postsynaptic potentials) (see Eccles, 1966; Shapovalov, 1966).

Repetitive responses to a steady current are also characteristic features of many nerve fibers, particularly sensory (Blair and Erlanger, 1936; Granit and Scoglund, 1943), for which a prolonged generator potential arising in a receptor is the natural stimulus (see the survey by Il'inskii, 1967).

The examination of the general principles governing repetitive responses is best combined with analysis of the mechanisms of production of this type of rhythmic activity.

We now know that generation of every action potential in response to application of a steady current is preceded by passive (electrotonic potential) and subthreshold active (local response) depolarization of the membrane. The spike arises as soon as the net potential change reaches the critical value.

Passive depolarization of the membrane is due to a fall in voltage on the resistance of the membrane (R_m), and since the resistance of the membrane falls sharply during the ascending phase of the action potential, the value of $I_m \cdot R_m$ during this period is also sharply reduced.

With recovery of the resistance of the membrane in the descending phase of the spike and in the period of after-hyperpolarization, if such takes place, the passive change of potential begins to increase and tends toward its initial prespike value.

The subsequent course of events depends on three conditions: 1) how quickly and how fully the resistance of the membrane is restored after the first spike; 2) whether the ability of the membrane to generate a local response of the previous magnitude is restored under these circumstances; 3) whether the initial critical level of depolarization is restored under these circumstances.

If all these properties of the membrane are restored sufficiently completely after the end of the first spike, in response to a current of rheobase or slightly higher strength the growing electrotonic potential will generate a local response which takes the depolarization of the membrane up to the critical level; a second spike arises, after which all the events are repeated in the same order.

However, this picture is found only in some excitable structures. It was observed by Hodgkin (1948) in axons (group I) of crustaceans and by Creutzfeldt et al. (1964) in the pyramidal neurons of the cat's cerebral cortex.

In most fibers and cells, however, after the end of the action potential the membrane resistance remains low (delayed rectification, connected with an increase in P_K), while the threshold of depolarization remains high because of partial inactivation of the sodium system.

Should these changes be small, they can be compensated by an increase in the strength of the stimulating current. In such cases, in response to application of a steady current of above-rheobase strength, repetitive discharges of spikes will take place.

If, on the other hand, the sodium inactivation of the membrane is strong, no repetitive discharge arises whatever the strength of the current applied. This happens in nerve fibers which have been damaged during dissection, depolarized by an excess of K^+ ions in the solution, exposed to the action of local anesthetics, and so on. Hyperpolarization of the membrane of fibers with low h_∞ by a steady current eliminates inactivation and restores their ability

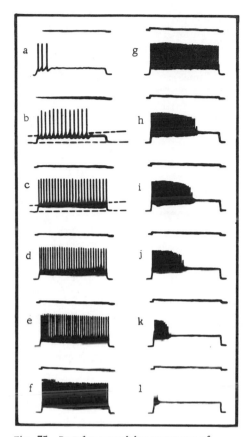

Fig. 75. Regular repetitive responses of
a crab nerve fiber to a steady depolarizing current 300 msec in duration. Strength
of current (in tests from a to l) increases
from just above threshold (10^{-8} A) to 6.5 x
10^{-8} A (Tomita and Wright, 1965).

to give repetitive responses to stimulation (Blair and Erlanger,
1936; Kvasov, 1937; Wright and Tomita, 1965).

However, even in nerve fibers in which inactivation of the
sodium system at the beginning of the steady current is small, it
increases progressively during depolarization. This leads to a
gradual increase in the critical potential and decrease in amplitude of the action potentials (Fig. 75). The stronger the depolarizing current, the more marked both these changes become, so that

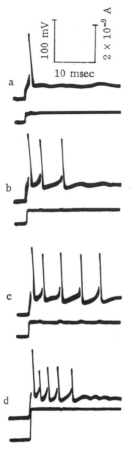

Fig. 76. Responses of a single node of Ranvier from an isolated frog nerve fiber to steady depolarizing currents of different strengths (a-d). Top line shows changes in membrane potential, bottom line strength of current. A decrease in the frequency of the discharge during depolarization (adaptation) takes place although the critical level remains unchanged (Vallbo, 1964b).

an increase in the strength of the current merely prolongs the discharge. If depolarization is increased excessively the duration of the discharge is sharply reduced because of the rapid development of cathodal depression (see Figs. 75 and 76).

The progressive increase in inactivation during repetitive activity is opposed by after-hyperpolarization of the membrane accompanying the spikes. It speeds recovery of the critical potential and thus facilitates the generation of a more normal response the next time (Tomita and Wright, 1965).

This evidently explains the fact that motoneurons in which after-hyperpolarization is of high amplitude and long duration are

able to give very long discharges (lasting many seconds) of spikes
in response to a depolarizing current of optimum strength.

This dependence of repetitive responses on the initial level
of inactivation of the sodium system explains the long familiar
association between accommodation of excitable structures and
their ability to give repetitive discharges of spikes in response to
steady stimulation (Katz, 1936, 1939; Sato, 1951).

Repetitive responses are known to arise to weaker currents
(relative to the threshold) and to continue longer in those excitable
structures in which accommodation is weak. For instance, repetitive responses are more marked in sensory fibers than in motor
fibers (see above); they are more marked in the internourons of
the spinal cord than in motoneurons (Shapovalov, 1966), and in the
body of the motoneuron than in its axon (Kernell, 1965).

In frog medullated nerve fibers the logarithm of the minimal
gradient is an almost linear function of the maximum number of
spikes in a repetitive series (Vallbo, 1964). If the minimal gradient exceeds 100 rheobase units/sec, no repetitive responses
will arise.

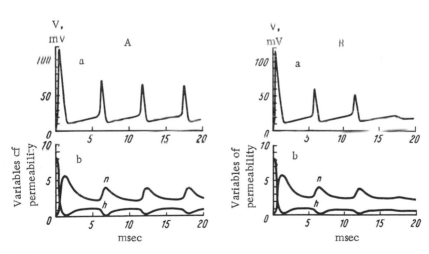

Fig. 77. Repetitive response obtained on a model of the membrane of the node of
Ranvier. a) Change in membrane potential; b) changes in constants of potassium
activation n and sodium activation h. A) For h = 0.8249, stimulation of steady current I_m = 0.6 mA/cm^2; B) for h = 0.8089, current 0.65 mA/cm^2 (Frankenhaeuser
and Vallbo, 1965).

All procedures increasing accommodation inhibit repetitive responses: they increase the threshold of their generation and shorten the duration of the discharge. Conversely, weakening of accommodation during hyperpolarization of the membrane facilitates the generation of long repetitive discharges to subsequent depolarization.

This dependence of repetitive responses on the initial level of inactivation and, consequently, on the rate of accommodation is clearly manifested when these phenomena are investigated on a mathematical model of the excitable membrane.

Frankenhaeuser and Vallbo (1965) showed that the model of the node of Ranvier gives repetitive responses to application of a steady current only if the initial inactivation of the sodium system is weak. For instance, with an initial value of $h_\infty = 0.8249$, a steady current evokes a virtually infinite discharge of spikes (Fig. 77A,a). However, only a very slight decrease in the value of h_∞ to 0.8089 (by changing the constant α_h) is sufficient to make the discharge finite in duration (Fig. 77B). With a decrease to 0.7728, however, the repetitive discharge completely disappears and only a single action potential is generated in response to application of the steady current.

Quantitative analysis of the relationships between the initial level of inactivation and the duration of the repetitive discharge was carried out experimentally by Bergman (1969). He showed that the number of spikes (n) per discharge in various sensory fibers of Rana esculenta and Xenopus is approximately proportional to the initial value of h_∞ at the resting potential.

Similar relationships were found also in experiments on the same fiber when h_∞ of the node of Ranvier was varied between 0.5 and 1.0 by hyperpolarization or depolarization of its membrane.

The investigations of Bergman (1969) also showed that, as well as a decrease in h, an increase in P_K during prolonged depolarization of the membrane also plays an important role in the limitation of the repetitive response. This is shown particularly convincingly by the fact that the addition of 5 mM tetraethylammonium to the Ringer's solution bathing the node considerably prolonged the repetitive discharge in sensory fibers (Fig. 78a) and led to the appearance of a repetitive response

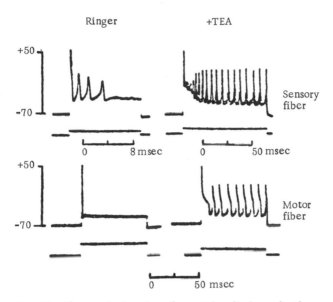

Fig. 78. Changes in duration of repetitive discharge by the action of TEA: a) maximal repetitive discharge of sensory fiber (single node of Ranvier) in ordinary Ringer's solution (left) and intensification of this discharge by 5 mM TEA (right); b) single response of motor fiber to steady current of above rhoobase strength in ordinary Ringer's solution (left) and appearance of a repetitive response in the presence of TEA (Bergman, 1969).

in some motor fibers (Fig. 78b; see also Fig. 51). This last effect, however, was observed only if h_∞ of the motor fiber was sufficiently high.

In Bergman's opinion, inability of motor fibers to give repetitive responses can be explained by the fact that the values of h_∞ and φ_K are lower in motor than in sensory fibers (see page 181).

The frequency of the spikes in a discharge depends on the rate of increase of depolarization in the interval between the spikes. Within certain limits this rate and, consequently, the frequency of the spike discharge are directly proportional to the strength of the current applied [as shown by the results of Granit, Kernell, and Shortuess (1963), Shapovalov (1964, 1965, 1966), and Kernell (1965) on motoneurons of the cat's spinal cord, by Creutzfeldt, Lux, and Nacimiento (1964) on cortical pyramidal neurons, and by Kandel (1964) on hypothalamic neuroendocrine cells, etc.].

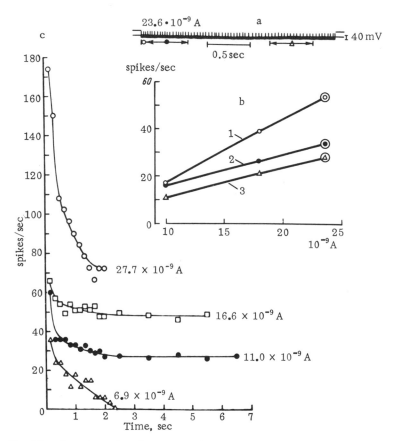

Fig. 79. The adaptation phenomenon in a motoneuron: a) repetitive
discharge of spikes in a cat motoneuron in response to stimulation through
an intracellular microelectrode; b) frequency of discharge (ordinate,
spikes/sec) as a function of current strength (abscissa, 10^{-9} A). Frequency measured as the reciprocal of the interspike interval. 1) 2nd-
3rd intervals; 2) 0.5 sec after 3rd interval; 3) 1.5-2 sec later (Kernell,
1965); c) frequency of spike discharge in motoneuron of cat's spinal cord
as a function of duration of action of the depolarizing current (abscissa,
in seconds). Strength of current shown on curves (Granit, Kernell, and
Shortness, 1963).

The curve showing the frequency of the discharges f as a
function of current strength I for spinal motoneurons remains
linear in character throughout the period of action of the current.
However, in the initial period the slope of the f versus I curve is
higher (Fig. 79b).

The range of frequencies within which the motoneuron gives stable discharges bears a close relationship to the duration of after-hyperpolarization. In motoneurons with short (about 50 msec) after-hyperpolarization this range is from 20 to 200 spikes/sec, while in motoneurons with long after-hyperpolarization (100 msec or more) the range is shifted toward lower frequencies (10-100/sec).

In spinal cord interneurons, because of the absence of after-hyperpolarization and the low level of accommodation, the maximum frequency of the repetitive response and the slope of the f versus I curve are higher than in motoneurons (Shapovalov, 1964).

Many interneurons, Renshaw cells, for example, are able to respond by a high-frequency discharge of up to a 1000-1500/sec to a single afferent stimulus evoking a long (of the order of 50-100 msec) EPSP (Eccles, 1959).

In the cortical pyramidal neurons the maximum frequency of the discharge (up to 450/sec) and the slope of the f versus I curve are higher than in motoneurons. The slope of the f versus I curve remains unchanged throughout the time of action of the current (Creutzfeldt, Lux, and Nacimiento, 1964).

In all the above-mentioned cells, just as in most (but not all!) nerve fibers, adaptation is a characteristic feature. This phenomenon is manifested by the fact that during the action of a depolarizing current of constant strength the length of the intervals between spikes increases progressively. At first, the frequency falls off rapidly (the initial period of adaptation), and then for a short time the frequency remains at a more or less stationary level; this is followed by a further slow decrease in frequency (the late period of adaptation) (Fig. 79c).

The results of investigation of repetitive discharges in single nodes of Ranvier and in motoneurons and interneurons of the spinal cord indicate that lengthening of the interspike intervals is not connected with an accommodative increase in the critical potential. The critical potential, as Fig. 76 shows, can be maintained at about the same level; only the rate of prespike depolarization is reduced.

To explain the mechanism of adaptation, it is natural to turn to mathematical models of the excitable membrane. However, neither the model of the squid giant axon nor the model of the node of Ranvier (Fig. 77) can reproduce the phenomena of adaptation

(FitzHugh, 1961; Tomita and Wright, 1965; Frankenhaeuser and Vallbo, 1965).

A careful examination of the records shown in Fig. 76 reveals that each lengthening of the interspike interval is preceded by a definite decrease in the level to which the value of V falls after the end of the spike. When a steady current passes through the membrane this level is directly dependent on the degree of increase in P_K and, correspondingly, the decrease in the net resistance of the membrane. Similar relationships were observed in the writer's experiments on nodes of Ranvier of Rana temporaria; an increase in "after-hyperpolarization," followed by lengthening of the interspike interval, is also observed in the records illustrating the paper by Bergman (1969). The causal link between these phenomena is clear: other conditions being equal, the higher the value to which P_K rises during the spike, the lower V must fall after its end, and, consequently, the longer the time interval which is needed for a fresh rise of V to the previous critical level. The model could reproduce this phenomenon if the reactivity of the potassium system (determined by the rate constants α_n and β_n, see pages 130-132) did not remain unchanged during the repetitive response, but increased with each spike. In this case, the increase in the variable of potassium permeability n (Fig. 77) during generation of the second spike would be greater than is found in this calculation, the after-hyperpolarization would be strengthened, and, correspondingly, the interspike interval would be lengthened. This, in turn, would lead to a more complete recovery of h and, in consequence of this, to an increase in the amplitude of the next spike, as is observed experimentally.

The role of the potassium system in the phenomena of adaptation is also shown by the work of Connor and Stevens (1971) who used the voltage-clamp method to study neurons of marine mollusks. These workers found that g_K rises during the repetitive response, and then falls very slowly after the end of the response.

However, the phenomena of adaptation cannot be attributed entirely to changes in the potassium system, because lengthening of the interspike intervals during the repetitive response persists after application of tetraethylammonium to the membrane (Fig. 79b), although it develops much more slowly under these conditions.

Changes in the constants of sodium inactivation (Franken-haeuser, 1963a) and, possibly, also in the leak conductance are evidently also concerned in the development of adaptation phenomena. This problem evidently requires further study.

Chapter VIII

Cable Properties and Geometry of Excitable Structures and Parameters of Their Electrical Activity

So far we have examined the effects of stimulation of a uniform area of an excitable membrane.

However, this is an artificial situation, because during ordinary stimulation of a nerve fiber (simulating the conditions of its stimulation by the receptor generator potential or by a traveling impulse) the changes in potential along the membrane are non-uniform, and "local circuit currents" arise between the excited and neighboring resting areas. These local currents have a significant effect on the development of excitation at the point of application of the stimulus, and they facilitate the spread of the excitation along the fiber.

The Propagated Action Potential

The cable properties of nerve fibers have already been examined in Chapter III. These properties can be expressed quantitatively (Hodgkin and Huxley, 1952d) by the following equation:

$$\frac{1}{r_0 + r_i} \cdot \frac{\partial^2 V}{\partial x^2} = i_m = C \frac{\partial V}{\partial t} + i_i \, , \tag{54}$$

where V represents the displacement of the potential from its resting level; r_0 and r_i the resistances of the external medium and the axoplasm per unit length of fiber respectively; i_m and i_i

213

the membrane (local) and net ionic currents per unit length respectively; and x the distance along the axon from the point of application of the electrode.

The left side of this equation characterizes the current of the local circuit, while the right side characterizes the radial current flowing through the membrane at this point (Fig. 76).

If the axon is immersed in a large volume of fluid the value of r_0 becomes so small that it can be disregarded. If r_i is now expressed in terms of the specific resistance of the axoplasm (R_i), Eq. (54) can be written as follows:

$$\frac{a}{2R_i} \cdot \frac{\partial^2 V}{\partial x^2} = C_m \frac{\partial V}{\partial t} + I_i, \tag{54a}$$

where a is the radius of the fiber.

By replacing I_i by the sum of its components I_{Na}, I_K, and I_l, Hodgkin and Huxley (1952d) obtained a nonlinear partial differential equation describing changes in the membrane potential (V) of the axon as a function of the distance (x), time (t), and ionic currents:

$$\frac{a}{2R_i} \cdot \frac{\partial^2 V}{\partial x^2} = C \frac{\partial V}{\partial t} + \bar{g}_{Na} m^3 h (V - V_{Na}) + \bar{g}_K n^4 (V - V_K) + \bar{g}_l (V - V_l). \tag{55}$$

Hodgkin and Huxley (1952d) examined the special case of propagation of the action potential along a fiber at a constant velocity θ, so that they were able to transform Eq. (55) into

$$\frac{a}{2R_i \theta^2} \cdot \frac{d^2 V}{dt^2} = C \frac{dV}{dt} + \bar{g}_{Na} m^3 h (V - V_{Na}) + \bar{g}_K n^4 (V - V_K) + \bar{g}_l (V - V_l). \tag{56}$$

By a method of trial and error these workers found a suitable value of θ for which Eq. (56) obtained a bounded solution. In this case the parameters of the calculated propagated action potential were very close to those found experimentally.

Such calculations gave, in the first place, a clear explanation of the time lag between the changes in impedance of the fiber and changes in membrane potential discovered earlier by Cole and Curtis (1939); it is clear from Fig. 80a that the increase in potential takes place before the beginning of the decrease in impedance.

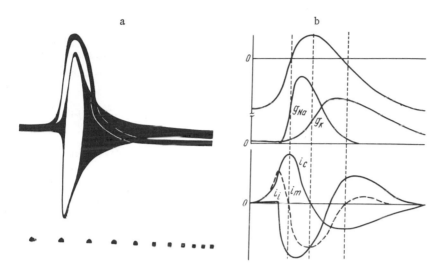

Fig. 80. Spreading action potential in a squid giant axon: a) change in impedance during propagated action potential (Cole and Curtis, 1939); b) changes in conductances and currents flowing through the membrane of the squid giant axon during a propagated action potential (schematic). Broken lines show corresponding points on the curves. Legend: i_i) net ionic current, I_m) local current; i_c) capacitance current (Noble, 1966).

The explanation of this phenomenon is that the initial part (the foot) of the action potential is the result primarily of passive depolarization of the membrane caused by the traveling impulse, i.e., by the positive local current I_m (see Fig. 80b)* which flows between the already excited area and the adjacent resting (less polarized) area of the membrane. It is only later that these passive changes of potential are joined by active changes connected with the increase in sodium conductance of the membrane and the inward sodium current (I_{Na}). It is then that the value of V begins to rise sharply and the impedance falls correspondingly.

During generation of a membrane action potential in response to a short stimulus, the peak of the spike corresponds to the moment when the net ionic current becomes equal to zero (see Fig. 30b).

---

*I_m and I_i are included in Kirchhof's equation (54) with different signs, so that a positive I_m, unlike a positive I_i, depolarizes the membrane, while a negative I_m repolarizes it.

The situation is completely different in the case of a propagated action potential. At the peak of the spike (Fig. 80b) the net ionic current I_i remains inward in direction and still possesses consider- able magnitude. This indicates that at that moment $I_{Na} > I_K + I_l$. Nevertheless, the action potential begins to fall.

The reason for the phenomenon we have just examined is as follows.

During the membrane action potential repolarization of the membrane can be brought about only by the outward I_i during the propagating spike; on the other hand, the local current (the nega- tive i_m in Fig. 80b), arising from less depolarized neighboring parts of the membrane, also participates in repolarization of the membrane. This local current stops the growth of the spike at a time when I_i is still inward in direction. Therefore, the peak of the propagated action potential corresponds to the time when $I_m = I_i$. This intervention of I_m explains the fact that the ampli- tude of the spike in the whole fiber (90.5 mV in the model) is lower than in a uniform membrane ($\sim$96 mV).

It follows from the above analysis that during the propagating action potential not only the beginning of the depolarization phase, but also the beginning of repolarization, is connected with currents of local circuits, i.e., they are passive in nature.

However, it must be emphasized that the role of the local current in the repolarization mechanism is important only in the initial period of repolarization. Later, I_m decreases and is re- versed (Fig. 80b), when the leading role in the process of re- storation of the initial level of the membrane potential is taken over by the outward potassium current.

Threshold Conditions of Stimulation

During the action of a short stimulus, the threshold condition for action potential generation in the uniform membrane is equality between the inward (I_{Na}) and outward ($I_l + I_K$) ionic currents (see page 154). At that moment

$$\frac{dV}{dt} = \frac{1}{C} \cdot I_i = 0.$$

This condition is insufficient for generation of an action potential in a continuous axon (case of "point stimulation") because the development of regenerative depolarization is prevented not only by the outward ionic current, but also by the current of the local circuit generated by neighboring resting areas of the membrane.

Generation of a propagating action potential thus requires that I_{Na} be equal to the sum of the local current $\frac{a\,\partial^2 V}{2R_i\,\partial x^2}$ and the outward I_K. It is only in this case that passive repolarization of the membrane will cease and a further increase in V will begin. At that moment

$$\frac{dV}{dt} = 0 \quad \text{and} \quad \frac{a\,\partial^2 V}{2R_i\,\partial x^2} = I_i.$$

If the stimuli are long (a steady current), the relationships are somewhat different.

Continuous depolarization of the membrane by the stimulating current facilitates generation of a propagating spike, and all that is required for its appearance is, therefore, that the rate of increase of the inward I_{Na} be greater than the rate of increase of the sum of the outward ionic current and the passive current of the local circuit.

The Threshold Potential

Since a stronger inward sodium current is necessary for generation of a propagating action potential, the threshold potential measured at the point of stimulation in this case will also be greater in magnitude (approximately twice as high in the giant axon) than in the case of spike generation in a uniform area of membrane (the membrane action potential).

However, an increase in the strength of depolarization of the membrane under the electrode, in the case of a continuous fiber, means an increase in the potential change at neighboring points of the membrane. The difference between the effects of subthreshold, threshold, and suprathreshold stimuli is therefore determined not only by the degree of the changes in potential, ionic permeabilities, and ionic membrane currents, but also by the area in which all these changes take place. An increase in the area of membrane in which the net ionic current becomes inward favors spike generation, for the repolarizing effect of neighboring unexcited regions of the membrane is thereby weakened in the central region (Noble and Stein, 1966).

All these hypotheses can be tested quantitatively only by solving the partial differential equation (55).

An important step forward in this direction was made recently by Cooley and Dodge (1966) (see also FitzHugh, 1962).

These workers prepared a program for the approximate solution of this equation by digital computer. For this purpose the "axon" was divided over a distance of 5 cm into a hundred segments, the properties of each of which were expressed by equations of Hodgkin and Huxley (1952d). Under the simplest conditions of point stimulation of the fiber through an intracellular microelectrode the calculations were able to predict the threshold, shape, and amplitude of the action potentials not only in the region of the fiber where the spike propagates at constant velocity, but also actually at the point of application of the stimulus.

Examination of Fig. 81a shows that in response to suprathreshold stimulation the amplitude of the calculated action potential in the segment of the fiber actually stimulated is virtually the same as the amplitude of spikes "recorded" at a distance (compare the spikes at the points $x = 0$ and $x = 3$). The differences between these action potentials relate only to the beginning of the depolarization phase.

Cooley and Dodge found a different picture if the stimuli were strictly threshold in strength (Fig. 81b,c). In this case the latent period between the end of the stimulus and beginning of the calculated spike is much longer (the more precisely the threshold is chosen, the longer the latent period), and as a result the amplitude of the action potential at the point of stimulation is significantly reduced by comparison with its amplitude at distances of 1 and 2 cm from the electrode. The reason for these differences is inactivation of the sodium system during growth of the local response.

It is a fact of great theoretical interest that at a certain strength of stimulation ($10^{-6}\%$ (!) below the threshold level) the local re-

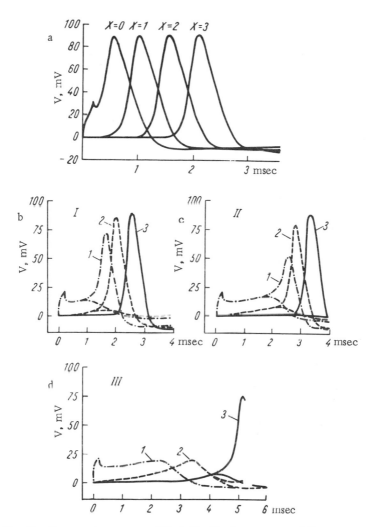

Fig. 81. Computed responses of Hodgkin-Huxley axon to a 0.2 msec long stimulus of ~1.5 threshold (10 μA). a) Propagating action potential at the points x = 0, 1, 2, and 3 cm from the point of stimulation (x = 0); b) responses of this axon to subthreshold and superthreshold stimuli of 0.2 msec duration. Accuracy with which the threshold was determined: 1% (I), 0.01% (II), 10^{-6}% (III). Curves 1, 2, and 3 show changes in V at the point of application of the electrode (x = 0) and at distances of 1 and 2 cm respectively from it (Cooley and Dodge, 1966).

sponse generated at the point of stimulation spreads over the first centimeter from the electrode as a subthreshold wave, and only then does it change into a normal action potential (see Fig. 81c). However, if the fluctuations of the resting potential, threshold, and ionic currents which are observed in the real axon (Blair and Erlanger, 1932; Verveen, 1961, 1962a–d; Derksen, 1965; Verveen and Derksen, 1968)* are taken into account, it becomes clear that the probability that such relationships are observed under normal physiological conditions is nil.

However, this probability increases significantly if the nerve fiber is anesthetized, i.e., when the inward sodium current is reduced. In this case the change from local response into action potential is hindered and the subthreshold wave, as it spreads along the fiber, becomes more stable.

The Threshold Current

The threshold strength of the current required to generate a membrane action potential, for a given value of the threshold potential (V_t), depends only on the resistance of the membrane itself (R_m). By contrast with this, in response to point stimulation of a continuous fiber, the voltage drop on the membrane depends, not on its R_m, but on the input resistance (R_{in}) of the fiber at this point (see page 39). The greater the value of R_{in}, the greater the change in potential produced by the current applied, and vice versa. In nerve cells with branching processes, and in nerve and muscle fibers with approximately identical threshold potential, the rheobase strength of the current is inversely proportional to their inward resistance.

For instance, R_{in} for pyramidal neurons of the cat's cortex, according to Creutzfeldt et al. (1964), is $(28 \pm 12) \times 10^6$ Ω, and rheobase current $(2.6 \pm 1.2) \times 10^{-10}$ A.

In motoneurons the inward resistance is $\sim 1.2 \times 10^6$ Ω and the rheobase $(5-8) \times 10^{-9}$ A. A low rheobase $(3 \times 10^{-10}$ A) is also characteristic of hippocampal pyramidal neurons, which also have a high inward resistance (about 13×10^6 Ω) (Kandel and Spencer, 1964).

---

*Poussart (1971) has recently investigated the nature of membrane fluctuations ("membrane noise") by the voltage-clamp method.

Other cells with a high R_{in} [$(11-50) \times 10^6$ Ω] and low rheobase $(5 \times 10^{-10}$ A) values are the neurosecretory neurons of the fish preoptic nucleus (Kandel, 1964).

What determines the input resistance of the cell?

In the case of spherical neurons with no dendrites, simple relationships exist between R_{in}, R_m, and the total membrane area S:

$$R_{in} = \frac{R_m}{S}.$$ (57)

The smaller the area of the membrane, i.e., the smaller the cell in size, the higher the value of R_{in}, and vice versa.

In nerve and muscle fibers the input resistance is directly proportional to the square root of the product of the resistances of the membrane (r_m) and of the axoplasm (r_i) per unit length of fiber:

$$R_{in} = \frac{1}{2} \sqrt{r_m r_i}.$$ (58)

Since

$$r_m = \frac{R_m}{\pi d}, \quad r_i = \frac{4R_i}{\pi d^2},$$

where d is the diameter of the fiber and R_i the specific resistance of the axoplasm,

$$R_{in} = \frac{1}{\pi} \sqrt{\frac{R_m R_i}{d^3}}.$$ (59)

This means that with a decrease in the diameter of the fiber the input resistance of the membrane rises considerably, and vice versa.

The fact that in nerve and muscle fibers the input resistance of the membrane is proportional to the square root of R_m, and not to R_m as in the case of spherical cells, is explained as follows.

As R_m increases the length constant λ [Eq. (12b)] rises, and together with it the dimensions of that area of the membrane through which the main part of the current flows increase correspondingly, thus partly compensating for the increase in R_m.

The fact that the threshold strength of the current is dependent on the passive electrical properties of the cell and, in particular, on its input resistance, is of great functional importance. The natural stimulus of the excitable membrane of the nerve cell and muscle fiber is in fact a synaptic current generated between the postsynaptic membrane, depolarized by ionic permeability changes controlled by the chemical transmitter, and the neighboring area of the electrically excitable membrane. Other conditions being equal, the higher the input resistance cell the greater the effectiveness of this current and, consequently, the greater the effectiveness of synaptic transmission.

Results so far obtained in fact indicate that the effectiveness of synapses located on small nerve cells, thin branches of dendrites, or thin muscle fibers, is much higher than that of synapses located on the bodies of large neurons or thick fibers.

The most convincing evidence in this direction is represented by the following facts obtained by Katz and Thesleff (1957) in their study of miniature potentials.

These workers found that the amplitude of miniature potentials in different frog skeletal muscle fibers varies by a factor greater than 10:1. The input resistance, which is inversely proportional to the diameter of the fiber, correlates well with the amplitude of the miniature postsynaptic potential: the thinner the fiber, the higher the value of R_{in} and the higher the miniature potential.

Assuming that miniature potentials of different magnitude are due to the liberation of equal portions of transmitter from the nerve ending, it can only be concluded that the effectiveness of the synaptic current depolarizing the membrane is determined by the shape and size of the cell.

Henneman, Somjen, and Carpenter (1965a,b), in experiments on decerebrate cats, determined the order of activation of various motoneurons in response to afferent nerve stimulation. They judged the size of the motoneurons indirectly from the amplitude of the spike recorded extracellularly from the ventral roots: the smaller the cell, the thinner the axon, and the lower the amplitude of the action potential shunted by the thicker fibers.

This investigation showed that, regardless of the method of activation of the neuron pool (ipsilateral or contralateral stimulation, adequate stimulation of receptors, or electrical stimulation of afferent fibers), or of whether flexor or extensor centers are excited under these conditions, the order of activation of the cells corresponds to their size. This result was obtained from 342 of 399 pairs of motoneurons.

On the basis of these results Henneman, Somjen, and Carpenter put forward the hypothesis that differences in the reflex excitability of motoneurons, the phasic or tonic character of the response, and even the character of differentiation of the muscle fibers innervated by them are largely determined by the size of these neurons and, consequently, by the magnitude of their input resistance.

Calculations by Arshavskii, Berkenblit, and Khodorov (1966) showed that the input resistance of terminal branches of dendrites, especially belonging to large neurons with long and highly branched processes (such as the motoneuron or pyramidal cell), is several tens of times greater than the input resistance of the cell body. This means that, when exposed to the same presynaptic activation, a postsynaptic potential tens of times larger must be generated in the terminal ramifications of the dendrites than in the soma of the neuron. The importance of this fact to the understanding of dendrite function will be obvious if it is considered that this strengthening of synaptic action can compensate for the reduced influence of postsynaptic potentials generated in branches of the dendrites on the cell body as the result of electrotonic decrement.

The Strength versus Duration Curve

Having solved the differential equation (55), Cooley and Dodge (1966) calculated the strength versus duration curve for a continuous axon. In principle, it was found to have the same shape as that previously calculated for a uniform area of membrane (see page 146) (Fig. 82a).

The differences between them, due to the cable properties of the fiber, were that at a given temperature the time constant of stimulation ($\tau_s = Q/I_0$) was approximately 30% shorter than for the space-clamp model, i.e., uniform area of membrane. In these workers' opinion, this effect is due to the fact that the continuous axon requires a somewhat larger depolarization of the membrane at the stimulating electrode for action potential generation in order to compensate for the electrical load of the adjacent passive

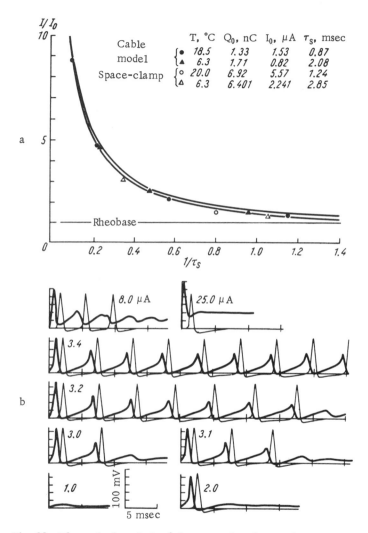

Fig. 82. Theoretical analysis of the properties of a continuous axon:
a) strength versus duration curves calculated for a continuous axon
(cable model) and for the space-clamped axon (uniform membrane).
Continuous lines represent limiting forms of the strength versus duration curves predicted by Hill's (1936) theory; b) repetitive responses
of a continuous axon to steady currents of various intensities. Changes
in membrane potential at the point of application of the electrode
(thick lines) and at a distance of 2 cm from it (thin lines) (Cooley and
Dodge, 1966).

cable; higher depolarization gives rise to more rapid ionic permeability changes underlying excitation.

It was also found that the sensitivity of the parameters Q_0 and
τ_s to changes in temperature were slightly higher in the cable model
than in the space-clamp model. The rheobase, however, was less
sensitive to temperature changes in the first case than in the second.

Repetitive Responses in a Continuous Axon

Repetitive discharges of action potentials in a continuous axon
are indistinguishable in principle from those recorded in a uniform area of membrane. Both types increase in frequency with
an increase in strength of the depolarizing current, and at the
same time the whole group discharge increases in length (the total
number of spikes in the discharge is increased). However, in
response to a very strong current, the group discharge rapidly
fails because of the development of cathodic depression.

Solution of Eq. (55) enabled Cooley and Dodge (1966) to
obtain repetitive responses for the cable model and to study the
differences between the changes in action potentials directly under
the electrode and some distance from it.

Examination of Fig. 82b shows that the amplitude of the responses at the point of application of the current has a tendency
to diminish during prolonged depolarization of the membrane. This
fall in amplitude of the spikes, connected with inactivation of the
sodium system, and the increase of g_K become more marked with
an increase in current intensity (see the effect of currents of 8
and 3.4 μA respectively). At a distance of 2 cm from the electrode, however, spikes of normal amplitude are "recorded" until
the local changes of potential fall below the threshold (for neighboring areas of the fiber) level.

Effects of Stimulation of Excitable
Structures with a Functionally
Nonuniform Membrane

The cable properties of nerve and muscle fibers enable the
electrotonic spread of depolarization or hyperpolarization created

by application of an electric current to one point of an excitable membrane.

If the active properties of this membrane are identical through-out the length of the fiber, excitation will arise at the point of application of the stimulus because depolarization is greatest at this point.

However, the excitability of the fiber can be reduced arti-ficially at the point of application of the stimulus, for example, by applying procaine solution in a concentration too low to block the nervous impulse completely to this area, when the following picture is observed: in response to a threshold depolarizing stimulus an action potential arises initially, not under the elec-trode, but nearby, in the more excitable neighboring area of the membrane from which it propagates both to parts of the axon un-affected by the procaine and also backward, into the altered area.

These phenomena, which can be observed in a nerve fiber under artificial conditions, simulate to some extent one of the most important mechanisms of normal activity of most nerve cells in the central nervous system of animals and man.

It is well known (Eccles 1959, 1966) that during synaptic stimulation of neurons an action potential is generated initially in the "trigger zone" (the initial segment), from which it propagates both along the axon and back into the body of the nerve cell.

This order of involvement of different parts of the neuron in the response is determined, in the modern view (Eccles, 1959), by differences in the excitability of the membrane of the initial segment and soma of the cell.

The threshold potential of the initial segment is taken as 7-10 mV, and that in the body of the nerve cell as 20-30 mV. Because of these differences, a weak postsynaptic potential spreading elec-trotonically from the soma of the neuron and its dendrites to the initial segment is sufficient to generate an action potential there sooner than in other parts of the cell. However, the initial seg-ment and body of the neuron differ not only in the values of their threshold potential, but also in other active properties of their membrane.

For instance, it has been shown for some motoneurons that the membrane of these parts of the cell respond differently to a slowly growing electric current.

According to results obtained by Araki and Otani (1959) on amphibian motoneurons and by Sasaki and Otani (1961) on cat motoneurons, the initial segment of many cells has a much higher mini-

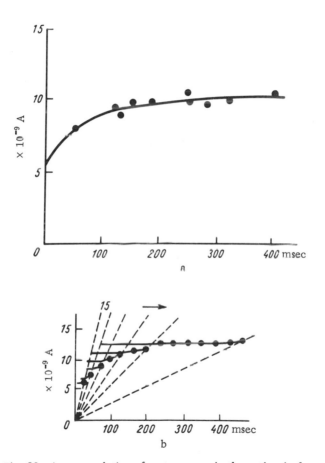

Fig. 83. Accommodation of motoneurons in the cat's spinal cord: a) abscissa, duration of linear increase in current before appearance of first spike, in msec; ordinate, strength of current at that moment. a) Accommodation curve of group I motoneurons; b) explanation in text (Sasaki and Otani, 1961).

mal gradient than the soma. During the action of rapidly increasing stimuli, an action potential thus arises in these neurons in the initial segment, while only the cell body responds to a slowly increasing current. Neurons of this type are characterized by an accommodation curve with a clearly defined plateau (Fig. 83a).

It is clear from this figure that slowing the increase in the current so that it extends over 200–300 msec doubles or trebles the threshold for these cells.

Since in response to local stimulation of the excitable membrane (for example, stimulation of a single node of Ranvier) by slowly increasing stimuli the threshold strength of the current is increased by more than 20–25%, these workers suggested that the accommodative increase in threshold observed on neurons under analogous conditions is only apparent, and is associated with a gradual migration of the point where the spike is generated from the initial segment to the soma of the cell. Each area of the neuron membrane, according to this hypothesis, has a minimal gradient which is highest in the low-threshold zone and which decreases gradually in the region of the cell body (Fig. 83b).

These examples show convincingly how important it is to consider the cable properties of excitable structures when the effects of their stimulation by stimuli of different strengths, duration, and steepness of growth are analyzed. At the same time, they evidently shed light on an important mechanism of differential response of different parts of the neuron to natural synaptic stimulation evoked by afferent impulses of increasing intensity and frequency.

The Relationship between the Velocity of Conduction of the Action Potential and the Cable Properties and Excitability of the Fiber

It can now be regarded as conclusively proved that the conduction of an action potential along a nerve or muscle fiber takes place with the aid of local currents flowing between excited and resting areas of the membrane (Tasaki, 1953; 1957; Hodgkin, 1965).

Investigations have shown that the velocity of conduction is proportional to the length constant of the fiber λ (see page 40). and inversely proportional to the time constant (τ_s) of the membrane (see the surveys by Chailakhyan, 1962, and Cole, 1968).

Since λ in nonmedullated nerve fibers is proportional to the square root of the diameter of the fiber [Eq. (12b)], the velocity of conduction, all other conditions being equal, is also proportional to $\sqrt{d}$ (Burrows et al., 1965).

In medullated nerve fibers, conduction is saltatory, i.e., it takes place from node to node. The length of the internode is approximately proportional to the diameter of the fiber, so that the conduction velocity in these fibers is proportional, not to $\sqrt{d}$, but to d (Rushton, 1951; Stämpfli, 1954; Kompaneets and Gurovich, 1966).

It is generally accepted that the conduction velocity depends on what is described as the safety factor, i.e., on the ratio between the amplitude of the propagating action potential and the threshold potential.

The relationship between the values of these parameters and the various constants of ionic permeability of the membrane was examined in detail in Chapter V. It was shown that the threshold potential is closely dependent on the reactivity of the sodium permeability system of the membrane. The higher this reactivity, i.e., the greater the increase in P_{Na} and, correspondingly, in I_{Na} for a given change of potential, the lower the threshold, and vice versa. Changes in the state of the potassium permeability system, as we have seen, have virtually no effect on the magnitude of the threshold potential. In precisely the same way, the leak conductance has very little effect on the threshold potential.

The relationship between the amplitude of the action potential of a membrane and the characteristics of its ionic permeability is much more complex in character. We have seen that an increase and decrease in the reactivity of the sodium system have little effect on the peak level of the spike (see page 120; Fig. 42b). The latter is significantly reduced only during strong inactivation of the sodium system (Fig. 33) or during a marked decrease in the maximum sodium permeability (Fig. 31).

Moderate changes in potassium permeability or in the reactivity of the potassium transport system have virtually no effect on the amplitude of the action potential (Fig. 46b).

Analysis of these results shows that, if the resting potential remains constant, the safety factor must increase if factors increasing the reactivity of the sodium system act on the nerve fiber (for example, with a decrease in the Ca^{++} ion concentration in the solution), and it must decrease during the action of factors reducing this reactivity (for example, an increase in the Ca^{++} concentration in the medium).

A very substantial decrease in the safety factor is produced by agents strengthening initial inactivation of the sodium system (decreasing h_∞) or decreasing $\overline{P}_{Na}$, so that in this case the spike amplitude falls while the threshold potential rises.

Changes in conduction of this nature were observed by Tasaki (1957) and others in response to the action of local and general anesthetics on the nerve fiber in concentrations too low to cause complete suppression of the action potential.

The level of the resting potential has a complex effect on the safety factor. Brief subliminal depolarization of the membrane, while not significantly altering the critical potential and amplitude of the action potential (see page 8), increases the safety factor, for it decreases the threshold potential ($V_t = E_c - E_r$). During strong depolarization, on the other hand, the spike amplitude falls and the critical potential rises, so that the safety factor is reduced (see page 9).

During hyperpolarization of the membrane the safety factor of the intact fiber falls, and the stronger the hyperpolarization, the greater the fall (since the threshold potential is increased under these circumstances). Should a hyperpolarizing current act on a membrane altered by procaine, cocaine, or ether, the amplitude of the action potential, which was reduced by these agents, increases but the increased threshold falls (see page 114). The safety factor therefore increases.

Besides the safety factor, the slope of the rising phase of the propagating action potential also has a significant effect on the velocity of conduction. This slope depends on both the passive and the active properties of the excitable membrane.

It was shown above that approximately the first third of the rising phase of the propagating spike is connected with passive depolarization of the nerve fiber membrane by the local current (see page 215). The rate of this depolarization for a given strength of local current is determined by the time constant of the membrane $\tau = R_m C_m$. The smaller this time constant, the faster the increase in depolarization and, consequently, the steeper the ascending phase of the spike.

The relationship between the slope of the rising phase and the active properties of the membrane was examined in Chapter V.

We saw that an increase in the reactivity of the sodium system shortens the rising phase, while a decrease in reactivity delays the rise of the action potential. Inactivation of the sodium system or a decrease in P_{Na} sharply reduces the rate of rise.

During the action of most factors, changes in the rate of rise of the action potential thus take place in the same direction as changes in the safety factor.

Only the time constant of sodium activation τ_m has a specific effect on the conduction velocity. In this case the safety factor is unchanged, for the amplitude of the spike and the threshold potential remain at the former level. On the other hand the rate of rise of the action potential falls with an increase in τ_m, but increases with a decrease in τ_m.

These changes in the rate of rise must explain the changes in the conduction velocity under the influence of cooling or heating the nerve fiber.

Conduction of Impulses along a

Nonuniform Fiber

Conduction of the action potential along a geometrically nonuniform excitable fiber is a problem which deserves special attention.

This problem was examined qualitatively in the investigations of Coombs, Curtis, and Eccles (1957), Fuortes, Frank, and Baker (1957), Arshavskii et al. (1966), Berkenblit et al. (1966), and Chailakhyan (1967).

The essence of these workers' arguments was as follows.

According to the local current theory, the amplitude of the propagating action potential (V_s), unlike that of the membrane action potential, depends not only on the emf of the excited membrane (E),* but also on the ratio between the internal resistance of the generator R_1 and the input resistance R_2 of the unexcited region of the fiber (load resistance):

$$V_s = E \frac{R_2}{R_1 + R_2}.$$

The higher the value of the ratio $R_2/(R_1 + R_2)$, the more the amplitude of the spreading spike approaches the value of E and, consequently, the higher the safety factor, and vice versa.

It thus follows, first, that a decrease in resistance of the membrane (an increase in its ionic conductance) during critical depolarization not only leads to spike formation, but also to an increase in the safety factor and, hence, an increase in the velocity of conduction of the impulse (Yamagiwa, 1956).

Second, it will be apparent that during the conduction of excitation along a geometrically nonuniform excitable structure the amplitude of the propagating spike must depend substantially on how close the area of fiber excited at a given moment is to the point of its branching or widening, and so on.

In fact, where a nerve fiber widens, for example, where it joins the body of the neuron or in the region of branching of the axon, the total cross-sectional area of the fibers and the total area of their membrane will be increased and, correspondingly, $R_2 = R_{in}$ will fall.

A decrease in R_2 reduces the safety factor and, correspondingly, the conduction velocity. Under certain conditions a decrease in R_2 may lead to complete blocking of conduction of the nervous impulse.

Phenomena of this type have been found experimentally at the point where the axon joins the cell body (Eccles, 1957; Ito, 1957; Lev, 1962; Shapovalov, 1966), at the point of branching of the axon

---

*E corresponds to the amplitude of the membrane action potential.

(Tauc, 1962a,b; Tauc and Hughes, 1963), and in the region of branching of muscle afferent fibers (Katz, 1950).

In the last case the block to conduction was observed only during orthodromic spread of excitation from thin terminal filaments to a thicker afferent fiber, i.e., in the direction of the greater drop of input resistance. During antidromic conduction of the impulse from the afferent fiber to the thin terminals, on the other hand, no block to conduction ever appeared.

These results and theoretical considerations permitted Arshavskii et al. (1966) to formulate an original hypothesis regarding the functional properties of dendrites which, in their opinion, are well equipped for the performance of various logical functions.

Such an investigation was undertaken by Khodorov, Timin, Vilenkin, and Gul'ko (1969) on a mathematical model of the squid giant axon similar to that used by Cooley and Dodge (1966). The only difference was that Cooley and Dodge investigated conduction of the nervous impulse along a fiber which was uniform throughout its extent, whereas the theoretical axon investigated by the present writer and his collaborators had the initial diameter (about 0.5 mm) only for the first 2 cm from the point of stimulation (i.e., from x = 0 to x = 2 cm), after which, starting from x = 2.025 cm, it widened suddenly by 3, 5, 6, or 10 times.

The calculations showed that the action potential passes easily through the threefold widening (Fig. 84a), passes with difficulty through the fivefold widening (Fig. 84b), and is completely blocked by the 5.5-fold widening.* The cause of the block is a sharp decrease in amplitude of the propagating spike near where the fiber widens. If the increase in diameter of the axon is insufficient to block conduction, as well as a decrease in amplitude of the action potential, characteristic changes also take place in the shape of its ascending phase. It will be clear from Fig. 84 that during a fivefold widening of the fiber in the segment x = 2 cm the increase in action potential takes place in two ways: initially V increases smoothly up to about 26 mV, and remains at this level for about 0.4 msec until a rapid rise of the spike begins in the next segment —

---

*Solution of this problem on an analytical model with simplified description of the properties of the membrane gave a lower critical value of widening, namely ≤ 4 (see Markin and Pastushenko, 1969; Chizmadzhev and Markin, 1970).

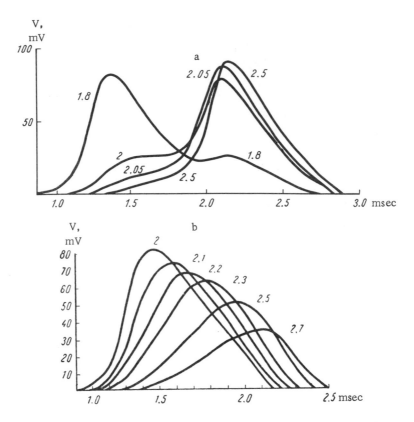

Fig. 84. Changes in calculated action potentials under the influence of threefold (a) and fivefold (b) widening of the axon. Numbers indicate values of x in cm (stimulation applied at the point x = 0 cm). Increase in diameter of fiber begins at x = 0.2025 cm (Khodorov et al., 1969).

the first widened segment (x = 2.05 cm). This at once speeds up the growth of potential in the preceding segment x = 2 cm, the depolarization wave spreads backward passively into the narrow part of the axon, and a hump is formed on the descending phase of the action potentials which have already terminated in this segment (see the curve for x = 1.8 cm).

Examination of the kinetics of the ionic and local currents enables a detailed analysis to be made of the mechanism of all these changes.

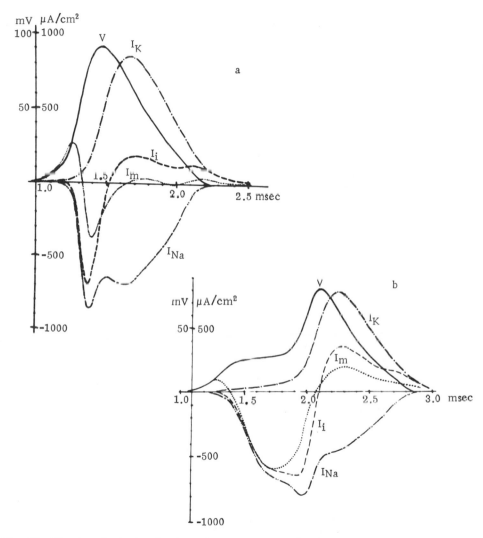

Fig. 85. Kinetics of calculated values of ionic (I_{Na}, I_K, I_i) and local (I_m) currents during generation of a spreading action potential at $x = 2.0$ cm. a) Homogeneous axon; b) axon with fivefold increase in diameter starting from point $x = 2.025$ cm.

It will be clear from Fig. 85b that, just as in the normal fiber (a), initial depolarization of the membrane at the point x = 2 cm is passive in character and is due to the positive I_m flowing in from previously excited areas. The inward I_i arising under the influence of this passive depolarization ought to give rise to a normal spike as in the case (a). This increase in V, however, is opposed by the repolarizing effect of the negative I_m arising from the still resting (less depolarized) regions of the axon ahead. Maintenance of the potential at the quasiplateau level corresponds to the period when I_m is approximately equal to I_i. Not until active depolarization begins in the widened part of the axon does I_m begin to decrease and to change its sign. This at once leads to a rapid increase of potential in the segment x = 2 cm. Most of this increase in V is active in character and is due to the inward I_i, and it is only near to the peak of the spike (starting from the time when I_m passes through zero) that a leading role in depolarization of the membrane comes to be played by I_m flowing in from depolarized segments of the thick fiber lying ahead.

Whereas at x = 2 cm the second increase in V is connected only partially with the positive I_m, in segments further from the point of widening of the thin fiber, at x = 1.8 cm, for example, the additional increase in potential (the hump on the descending phase of the spike) is completely and entirely determined by the positive local current I_m. It is clear from Fig. 86 that the net ionic current I_i at this time remains outward in direction, because the increase in depolarization of the membrane as a result of the low value of h cannot evoke a further substantial increase in the inward I_{Na}.

The conditions for conduction of impulses through points of branching are similar in many respects to the conditions for the spread of excitation into the region of widening of the fiber. In fact, just as in the case of widening, where the axon branches there is a substantial increase in the total area of the membrane and, consequently, a decrease in its net resistance per unit length and a corresponding increase in the capacitance of the membrane per unit length. As a result, the length constant λ and, consequently, the region of the membrane affected by the local current, are increased. The only differences between the conditions of conduction of nervous impulses through regions of widening or of branching are as follows. Where the fiber widens, as well as a de-

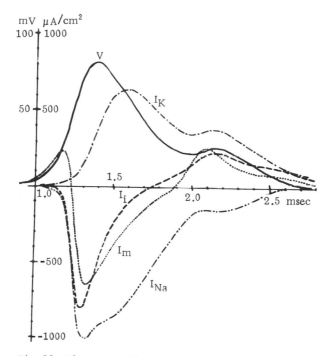

Fig. 86. The same as Fig. 85, but at the point $x = 1.8$ cm.

crease in resistance of the membrane, there is also a decrease in resistance of the axoplasm per unit length. In the case of branching, however, the total resistance of the axoplasm of the branches per unit length may be increased, reduced, or even unchanged depending on the ratio between the area of cross section of the main fiber and the combined areas of cross section of its branches. Clearly, if the latter is increased, conduction through the point of branching may be substantially disturbed.

It follows from the foregoing argument that the principles governing the spread of the impulse through a region of widening also are applicable, in principle, to a point of branching.

Strict evidence in support of this conclusion was obtained on an analytical model by Pastushenko, Markin, and Chizmadzhev (1969) and on a Hodgkin—Huxley model by Berkenblit et al. (1971).

Special investigations (Khodorov et al., 1971) have shown that the generation of retrograde depolarization waves during the conduction of impulses through a region of axon widening or branching is the main cause of the rhythm transformation in these areas. The retrograde wave in fact does not affect conduction of the particular spike concerned, but it severely impairs the propagation of the next spike through the nonuniformity if it follows only a relatively short time after the first. The reason for this phenomenon is understood: the retrograde wave lengthens the time of membrane depolarization and thus inhibits recovery of h and g_K after the end of the first action potential (Fig. 87). As a result, as the second spike approaches the point of axon widening, it finds each successive point of the fiber in a deeper state of refractoriness than the preceding point. This leads either to complete blocking of the second spike (if the axon widens considerably and the interspike interval is short, as in Figs. 87 and 88), or to an increase in the delay of its conduction into the widened region. How-

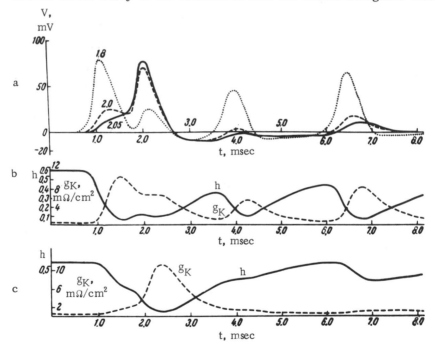

Fig. 87. Changes in potential V, potassium conductance g_K, and inactivation variable h during conduction of a series of impulses into the region of a fivefold widening of an axon. Diameter of fiber starts to increase at the point x = 2.025 cm. a) Changes in V at the points x = 1.8, 2.0, and 2.05 cm. b) Changes in g_K and h at the point x = 1.8 cm. c) The same, at the point x = 2.0 cm. Interval between stimuli 2.5 msec.

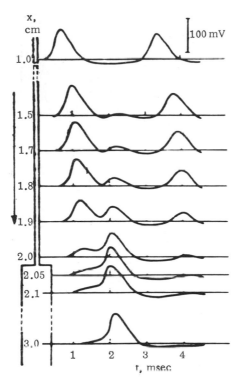

Fig. 88. Retrograde wave of depolarization
and its effect on conduction of a subsequent
spike into the region of widening. Changes
in V are shown at different points of the
axon during conduction of two spikes at an
interval of 2.5 msec. The axon is shown
schematically on the left. Numbers correspond to values of x in cm (Khodorov et al.,
1971).

ever, lengthening the delay strengths the retrograde wave, and
this impairs conduction of the third spike still more, and so on.
As a result of this cumulation of delays and enhancement of the
retrograde waves, one of the spikes in the series will be blocked,
after which conduction will be restored, and so on.

A good example of this periodic blocking of impulses (Wenckebach's period) is illustrated in Fig. 89. In this case the first four
spikes pass the boundary of the widened region with increasing
delay, while the fifth spike, after the powerful retrograde wave

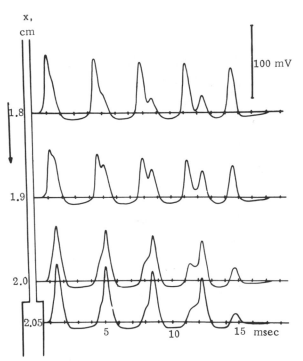

Fig. 89. Wenckebach's period for conduction of a series
of spikes into a region of threefold widening of an axon.
Interval between stimuli 3.3 msec (Khodorov and Timin,
1971).

accompanying the fourth spike, was omitted; thereafter the block
recurred after every three spreading action potentials. The way
in which the rhythm transformation depends on the coefficient of
fiber widening and the duration of the interspike interval is shown
in Table 3.

This table shows how even comparatively small changes in
the constants of ionic permeability of the membrane in the region
of axon widening can have a powerful effect on the character of
rhythm transformation.

These results, which were all obtained by the use of the mathematical model, are important for the correct interpretation of
those distinctive changes in the shape and amplitude of action potentials observed experimentally during the antidromic conduction
of excitation from the initial segment into the cell body (Coombs,

TABLE 3. Type of Transformation in Relation to Character of Nonuniformity

Coefficient of fiber widening	Additional factor	Interspike interval (msec)			
		2.5	3	3.3	3.5
		number of blocked impulses			
1.5	—	—	—	—	—
3	—	2, 4, 6, ...	3, 5, 7, ...	5, 9, ...	
3	$\Delta B = 3$ mV		2, 4, 6, ...		
3	$\Delta B = -5$ mV		4, 8, 12, ...		
3	$\bar{g}_{Na}/2, \bar{g}_K/2$			3, 5, 7, ...	
5	—	2, 3, 4, 5, ...	2, 3, 4, 5, ...	2, 3, 4, 5, ...	2, 3, 4, 5, ...
6	—		1, 2, 3, 4, ...	1, 2, 3, 4, 5, ...	1, 2, 3, 4, ...

N o t e . ΔB represents the change in the constant B in Eqs. (41)-(48) from its initial value (see Table 1). An increase in B corresponds to an increase in the outer Ca ion concentration, and a decrease in B to a decrease in $[Ca]_0$.

Curtis, and Eccles, 1957; Eccles, 1957; Fuortes, Frank, and Baker, 1957).

It will be seen in Fig. 90 how spikes with double peaks, very similar to those obtained with the model near the place of widening of an axon, are recorded when the microelectrode lies in the initial segment of a motoneuron. When, however, potentials are re-

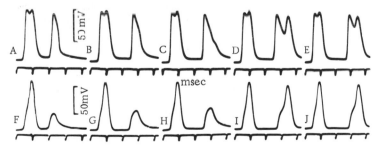

Fig. 90. Intracellular responses evoked by impulse arising in a motor axon in response to application of two stimuli at different intervals to the nerve to the biceps femoris muscle. Resting potential −82 mV. It is assumed that in A-E the microelectrode tip was in the region of the initial segment of the axon, whereas in F-J it was moved into the soma of the motoneuron. Experiments on cat (Coombs, Curtis, and Eccles, 1957).

corded from the cell body (Fig. 90, F-J) the potential rises in two steps, in a manner similar to that obtained by calculation in the first segments of the widened part of the fiber (at x = 2.05 and 2.1 cm on the curve of the changes in V in Figs. 87 and 90).

The similarity between the experimental and calculated results is emphasized also by the characteristic way in which the shape of the second spike depends on the interspike interval: if this interval is short the second spike cannot pass into the cell body, and the retrograde wave is then absent (Fig. 90A-B); prolonging the interval restores conduction, and the second peak then appears (Fig. 90C-E).

Results of the calculations provide an unequivocal interpretation of curves like those shown in Fig. 90A-E. A propagating spike with a double peak can arise only if the amplitude of the spike is reduced actually at the point of recording, and it is conducted with delay into the segment in which the amplitude of the response is high. In this case, the duration of the delay can be judged from the distance between the two peaks at a certain point x, and correspondingly, the difference between V_s at that point and at preceding points of the fiber (if the distance between them is known), or the distance between them (if the amplitude of the responses of these points is known) can be judged from the amplitude of the second peak.

The question may be asked how the action potential is conducted from the initial segment into the neuron soma if the difference between their diameters is significantly greater than the critical widening (5-6) obtained by calculation. The reason for this apparent disagreement is that in the model the widening conventionally took place in a step, whereas in the neuron the transition from nonmedullated axon to axon hillock and beyond, into the soma, is gradual. The significance of this factor has been studied in detail on the Hodgkin—Huxley model by Berkenblit et al. (1970). Further investigations showed that a retrograde wave of depolarization arises not only during conduction of an impulse along a geometrically nonuniform fiber, but also if certain functional nonuniformities exist in the axon (Khodorov and Timin, 1970, 1971a,b). In this case also, the condition for appearance of the retrograde wave is inequality of amplitudes of the action potentials on the two sides of the boundary of the nonuniformity. For instance, if

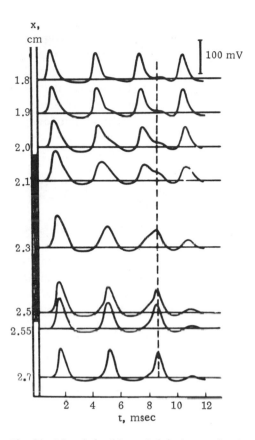

Fig. 91. Wenckebach's period during conduction
of a series of impulses through a cooled segment
of an axon. Cooling from 20 to 16°C simulated
by doubling all the time constants [$\tau_m(V)$, $\tau_n(V)$,
$\tau_h(V)$] in the segment of fiber from $x = 2.025$ to
$x = 2.525$ cm (the shaded region). A broken ver-
tical line is drawn through the peak of the retrograde wave of depolarization causing blocking of
the fourth spike. Interval between stimuli 3
msec (Khodorov and Timin, 1970).

the amplitude of the action potential is reduced in a certain segment of the fiber (for example, through a decrease in $\overline{g}_{Na}$) on
passage into the normal zone the high-amplitude spike developing
there will become the source of a retrograde depolarization wave
which will spread into the altered segment. The principles gov-

erning changes in retrograde waves during conduction of a series of spikes through a functionally nonuniform region are basically similar to those obtaining for geometrical nonuniformity. In both cases the cause of the periodic blocking of the spikes (Wenckebach's period) is prolongation of the refractory state due to the generation of a retrograde wave of depolarization as the excitation passes across the boundary of the nonuniformity. As an example, blocking of the fourth spike during the conduction of a series of spikes through the "cooled" segment of a theoretical axon is illustrated in Fig. 91. The cause of the block in this case is the appearance of a retrograde wave as the third spike passes the distal boundary of the altered segment.

It must be emphasized that in the case of "cooling" the retrograde wave depolarization arises not only at the exit, but also at the entrance into an altered zone. In the case represented in Fig. 92a, the marked lengthening of the descending phase of the action potential in the "cooled" segment delayed repolarization of the membrane in the "warm" zone of the fiber before the boundary of the "cooled" segment (see the changes in shape of the spikes at points $x = 1.8$ cm, 1.9 cm, and 2.0 cm). The changes in ionic and local currents taking place under these circumstances are shown in Fig. 92b.

Special attention must be paid to the analysis of changes in amplitude of the propagating spike (V_s) in a functionally nonuniform fiber. Investigations on mathematical models of the nonmedullated (Khodorov et al., 1970; Khodorov and Timin, 1971a) and medullated (Khodorov and Timin, 1971b) fibers showed, first, that V_s undergoes regular changes not only in the altered segment itself, but also just before it reaches that segment. Regardless of the method used to reduce excitability in the altered segment, V_s begins to fall under the influence of the increased negative I_m close to its proximal boundary.*

In the uniform fiber I_m becomes negative in sign (i.e., becomes repolarizing) only for a short time at each point, for the preceding resting segments of the fiber, which are the source of the current, themselves pass rapidly into a state of excitation

---

*With the mathematical model this boundary can be made as abrupt as desired, but for reasons which can be understood, this is extremely difficult to achieve in experiments on living axons.

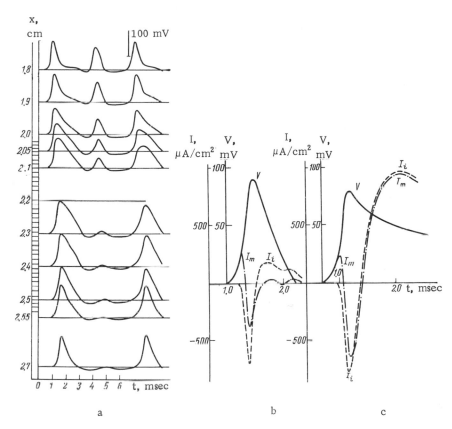

Fig. 92. Conduction of impulses through a segment of an axon cooled from 20 to 10°C [simulated by a threefold increase in the time constants $\tau_m(V)$, $\tau_n(V)$, $\tau_h(V)$]: a) Legend as in previous figures; b and c) comparison of kinetics of the net ionic current I_i and local current I_m during spike generation at the point $x = 2.0$ cm in a normal axon at 20°C (b) and in an axon whose segment starting from $x = 2.025$ cm is cooled from 20 to 10°C (c). Note the decrease in amplitude of the spike due to an increase in the negative I_m in (c).

(Fig. 93a). When the spike comes close to the region in which, because of reduced excitability, spike generation is delayed or completely impossible, the repolarizing I_m is increased in strength and duration (Fig. 93b), and this naturally leads to a decrease in both amplitude and rate of rise of the action potential.

Changes in the action potential and local current before a region with reduced excitability are thus similar, in principle, to those observed close to a point of widening or branching of a fiber.

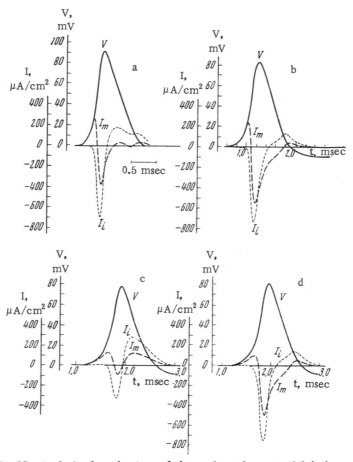

Fig. 93. Analysis of mechanisms of change in action potential during its conduction through a segment of an axon with 50% decrease in $\overline{g}_{Na}$: a) action potential, I_i and I_m at the point x = 2.0 cm of a normal axon; b) the same, but in an axon for which the value of $\overline{g}_{Na}$ is reduced by 50% for the segment from x = 2.025 to x = 3.025 cm; c) changes in action potential and currents I_m and I_i in the last segment of the altered region (x = 3.0 cm); d) action potential and currents in the first segment of the normal zone (x = 3.0 cm). Explanation in text (Khodorov and Timin, 1971a).

Calculations showed that the local currents have a very considerable influence on spike amplitude in the altered region also.

Changes in amplitude of action potentials during their passage through a segment of fiber in which the maximum sodium conduc-

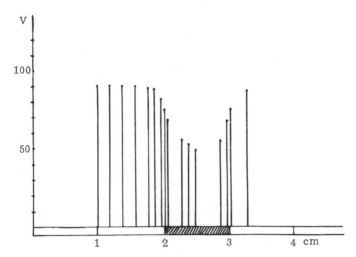

Fig. 94. Changes in amplitude of calculated action potentials at different points of an axon for which the value of $\overline{g}_{Na}$ is reduced by 50% in the segment from $x = 2.025$ to $x = 3.025$ cm. The altered zone is shaded. Abscissa: values of x, in cm; ordinate: values of V, in mV.

tance $\overline{g}_{Na}$ is reduced to half its standard value (simulation of the tetrodotoxin effect) are shown in Fig. 94.

As the spike moves further into the altered segment, V_s at first decreases, but near to the distal boundary of this region it once again begins to increase gradually. This behavior is due primarily to changes in the properties of the action potential associated with the changes in the properties of the membrane mentioned above. In the giant axon a decrease in $\overline{g}_{Na}$, if the potassium conductance of the membrane remains unchanged, makes the spike extremely gradual, so that it does not obey the "all or nothing" rule.*

It is also useful to examine the causes of gradual restoration of the amplitude of the spike after its departure from a zone of lowered $\overline{g}_{Na}$.

---

*Theoretically, the amplitude of the action potential in a giant axon at the point of stimulation is always dependent on stimulus strength I. However, under normal conditions the I versus V_s curve is very steep, so that the gradual nature of the spike can be detected only by the use of very finely graduated strengths of the stimulating current, a virtual impossibility because of continuous fluctuations of the threshold (see page 218; see also: Cole, Guttman, and Bezanilla, 1970). If $\overline{g}_{Na}$ decreases, the gradient of the I versus V_s curve falls sharply and the gradual character of the spike becomes easily observable.

The calculated action potential at a point x = 3.05 cm is shown in Fig. 93d. Its amplitude (82.5 mV) is about 11% below its initial level, despite the fact that $\overline{g}_{Na}$ has the normal value in this segment and the inward net ionic current I_i (750 $\mu A/cm^2$) is actually slightly higher than the maximum of I_i in the normal axon (700 $\mu A/cm^2$, see Fig. 93a). This relative decrease in the spike at x = 3.05 cm is due to an increase in the strength and duration of the local repolarizing current (phase II of I_m), and it is outwardly very similar to that taking place during action potential development at x = 2.0 cm (compare I_m in Fig. 93b and d). The only difference is that at x = 2.0 cm the sources of the local repolarizing current are the segments of the fiber lying ahead with reduced $\overline{g}_{Na}$, while at x = 3.05 cm the repolarizing I_m is strengthened on account of the altered segments which now lie behind.

This account explains why changes in the gradual spike during its conduction along an altered segment must be closely dependent on the length of the segment. If the gradual character is very conspicuous and the length of the altered segment is great, the spike must inevitably die away before reaching the normal zone.

This proposition is theoretically applicable not only to nonmedullated, but also to medullated nerve fibers (Khodorov and Timin, 1971b).

Some Properties of Spike Conduction from a Medullated into a Nonmedullated Segment of a Nerve Fiber

Conduction of the impulse in this transitional zone meets the same difficulties, in principle, as the spread of excitation into a region of widening of an axon. In fact, in both cases there is an increase in the area of membrane affected by the local current in the nonuniform segment in both cases, and, consequently, the density of this current falls. This is equivalent to a decrease in the safety factor.

Special calculations by Timin and the author showed that an action potential arising in the last node could not, generally speaking, excite the nonmedullated part of the fiber unless there was a substantial decrease in diameter of the axon at the point of transition.

Narrowing of the fiber effectively reduces the length constant λ of the terminal, and this increases the density of the local
current and thus facilitates the transmission of excitation from
the last node.

Model investigations also showed that in the nonmedullated
part of the fiber there must also be a significant increase in R_m
compared with its value for the nodes of Ranvier. If the nonmedullated terminal had a membrane with an R_m as low as that of the Ranvier node (10–$30 \; \Omega \cdot cm^2$), then, other conditions being equal, the
spread of nerve impulse conduction would be considerably (more
than twice) lower than it really is. Computations have shown that
an increase in R_m of the theoretical terminal from 30 to 1500 $\Omega \cdot$
cm^2 (i.e., to the value of R_m of squid giant axon) raises the ratio
between the amplitude of the electrotonic potential in the first
segment of the terminal and the amplitude of the action potential
in the last node from 0.17 to 0.42.

Conduction into a nonmedullated terminal, as calculations
showed, can also be substantially facilitated by shortening the internodal segments (i.e., by a decrease in L/D, where L is the length
of the internodal segment and D the diameter of the medullated
fiber) near to the transition zone. The velocity of impulse propagation under these circumstances is, of course, reduced, but the
efficiency of transmission of excitation from the node into the
nonmedullated part of the fiber is substantially increased. The
morphometric findings of Zencker (1964) and others show that
this method of facilitation of the transmission of excitation into
nonmedullated terminals is evidently widely used in nature. For
example, in motor fibers innervating the external ocular muscles
of monkeys the value of L/D near the point of transition into nonmedullated terminals (these arise directly from nodes of Ranvier)
are reduced to 5–10, i.e., by 10–20 times compared with the values
of L/D found in the central parts of the axon.

However, it must be emphasized that, despite all these adaptive changes in the geometry and functional properties of the nerve
fiber where it changes into a nonmedullated terminal, under real
conditions this region as a rule has a comparatively low safety
factor.

This is seen, first, from the fact that conduction of the nervous impulse is greatly retarded in this zone. Braun and Schmidt

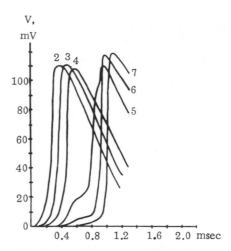

Fig. 95. Calculated action potentials in the last three
nodes of Ranvier (2-4) of the medullated part of a fiber
where it changes into the nonmedullated terminal; 5-7)
the first three segments of the nonmedullated terminal,
each 0.06 mm in length. Investigation on model formed
by combination of the Frankenhaeuser—Huxley (1964)
equations describing the properties of the membrane of
the Ranvier node, and cable equations describing the
properties of the medullated segments (see Khodorov and
Timin, 1971b). Diameter of terminal is 1/3 of the diameter of the axon in the medullated part of the fiber. Values of g_l and C_m for the terminal are reduced by 5 and
2 times respectively. Diameter of axon, 10 μ, L/D = 52.
The increase in spike amplitude in the nonmedullated
terminal is due to a decrease in g_l. Note the fall of the
spike in the last node, characteristic of conduction of an
impulse in a segment with lowered safety factor.

(1966), for instance, who studied nerve-muscle preparations of
frogs, found that over the first 30 μ of the nonmedullated terminal
the conduction velocity is only 14–18 cm/sec, whereas in the more
distal parts of the same terminal the velocity reaches values of the
order of 85–126 cm/sec.

During the action of various procedures reducing excitability
of the nerve fiber in this zone, blocking of conduction of the action potential often arises because of the low safety factor.

This evidently takes place in cases of asphyxia of the spinal
cord (Collwein and Van Harreveld, 1966), the action of barbiturates (Weakly, 1969), high-frequency stimulation of axons, and so on.

The blocking of antidromically propagating impulses in the region of transition between a medullated axon and the nonmedullated initial segment of the neuron observed under certain conditions is based on a similar mechanism (Coombs, Curtis, and Eccles, 1957; Lev, 1964).

Since intracellular recording of potentials from nonmedullated terminals in experiments on living objects is not yet a practical possibility, these changes were investigated on models. The results of one of the calculations is shown in Fig. 95. In the first segment of the nonmedullated terminal, the spike arises after a delay due to the slow increase of the potential up to the critical level. The direct cause of this delay is a decrease in the density of the local current I_m in the transitional region of the fiber.

The vulnerability of the zone of transition from a medullated fiber into a nonmedullated terminal is perhaps used by nature in the mechanism of presynaptic inhibition.

Compensatory Changes in Ionic Currents during Conduction of the Impulse along a Nonuniform Fiber

It was shown above that a decrease in the amplitude and maximum rate of rise of the action potential as it moves closer to the point of axon widening (Figs. 85 and 86) or to a segment where excitability is reduced by lowering $\overline{g}_{Na}$ or by cooling (Figs. 92b and 93b) is due to an increase in the negative (repolarizing) I_m.

If these figures are examined carefully, however, we see that there is a marked quantitative disproportion between the changes in I_m and the changes in V_s and $\dot{V}_{max}$; the increase in I_m is always greater than the decrease in V_s and $\dot{V}_{max}$.

For example, in the case of a fivefold widening of a theoretical axon, the peak of I_m at the point $x = 1.8$ cm (Fig. 86) is increased by about 70%, while V_s and $\dot{V}_{max}$ are reduced by only 12 and 15% respectively, compared with their original values in the uniform fiber (Fig. 85a).

These facts which at first sight appear paradoxical can be explained by the hypothesis that the changes in the kinetics of V resulting from the increased intensity of the local current I_m lead

to an automatic increase in the inward I_{Na} and, consequently, of I_i
In other words, in response to an increase in load the membrane
generator begins to develop additional power (Timin, 1971). This
increase in intensity of the inward ionic current is manifested as
an increase both in the peak of the negative I_i and a particularly
large increase in the value of I_i at the peak of the spike. For instance, in the example of a fivefold widening of an axon examined
above, the I_i peak at the point $x = 1.8$ cm was increased by about
16%, while the value of I_i at the peak was increased by 160% compared with their values in the uniform axon.

The mechanism of these changes in ionic current, which are
compensatory in character, is as follows.

During regenerative depolarization of the membrane, the value
of g_{Na} for each value of V depends on the rate with which V reached
this value: the slower the growth of V, down to a certain limit,
the larger the number of sodium channels capable of being activated
in that time, and vice versa (Fig. 96). Admittedly, besides an increase in the activation variable m, a decrease in h (i.e., the development of inactivation) can reduce the rate of increase of V, but
since $\tau_h \gg \tau_m$, the changes in h associated with moderate varia-

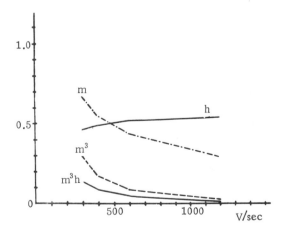

Fig. 96. Values of m, h, m^3, and m^3h at $V = 100$ mV
as functions of gradient with which V reaches this
level in the linear part of its growth. Calculated by
Eqs. (21)-(23) and (27)-(30), under the condition
$\dot{V} = $ const.

tions of the rate of rise of the action potential have little effect on the values of $g_{Na}(V) \sim m^3 h$.

Any deviations of $\dot{V}_{max}$ from its initial value (whether a decrease or increase) thus leads to a compensatory increase or decrease in the I_i peak, as a result of which this deviation is reduced.*

Stabilization of the amplitude of the action potential is based on different mechanism. It is determined mainly by the dependence of I_{Na} on the electrochemical potential $V - V_{Na}$. In the rising phase of the spike the value of $V - V_{Na}$ falls continuously, and this diminishes the growth of I_{Na}, giving the biphasic curve of I_{Na} (see pages 98-103). As a result of this decrease in I_{Na}, the current repolarizing the membrane ($I_K + I_l$ in a uniform membrane, $I_K + I_l - I_m$ in a whole fiber) becomes equal to the outward depolarizing current I_{Na}, and growth of the spike is therefore stopped.

However, the dependence of I_{Na} on $V - V_{Na}$ not only plays an important role in limiting the growth of the spike, but it also prevents its decrease on account of all the factors inhibiting regenerative depolarization of the membrane. Any tendency toward a decrease in amplitude of the action potential in fact leads to a relative increase in $V - V_{Na}$ for a given value of g_{Na}, and, consequently, it also leads to an increase in I_{Na}. As a result, the rate of decrease in amplitude of the spike is reduced.

The functional importance of this stabilization of the amplitude and rate of rise of the action potential is clear: it substantially increases the reliability of conduction of the impulse into a region of widening or branching of a fiber, at the junction between the medullated part of the fiber and its nonmedullated terminal,

---

*This rule is demonstrated very clearly in the following model experiment. Let C_m of the node of Ranvier be increased by 3 times. From Eq. (14a) it would be expected that $\dot{V}_{max}$ of the membrane action potential would be reduced under these circumstances by the same number of times. In fact, $\dot{V}_{max}$ falls under these conditions by only 1.5 times, since the slowing of the ascending phase of the spike doubled the inward I_i. In the squid giant axon a threefold increase in C_m is compensated to a lesser degree: $\dot{V}_{max}$ is halved. The explanation of this fact is that in the giant axon the chief component of the outward current is not I_l (as in the node), but I_K, and if the ascending phase of the spike is retarded, this increases, thereby reducing the net current I_i.

in a zone with reduced excitability, and so on. In particular, calculations have shown that if I_i did not increase before the region of widening of the fiber, the block to conduction would occur if the increase in diameter was only 3.5 times, and not 5.5 times (see the footnote to page 233).

An effective compensatory change in ionic currents is also observed during the falling phase of the action potential: any tendency toward an increase in the rate of repolarization of the membrane (for example, by a negative I_m close to the segment with reduced excitability; Fig. 93b) decreases I_K and increases I_{Na}, whereas a delay in repolarization, on the other hand, increases I_K and reduces I_{Na}.

As an example let us examine changes in ionic currents during the descending phase of the action potential at the point $x = 1.8$ cm in the case of a fivefold widening of the axon (Fig. 86). Delay in conduction of the impulse into the widened part of the axon leads to a significant increase in the positive I_m (phase III) which forms the additional rise of potential – the "hump" on the falling slope of the spike. It is easy to show that the amplitude of this hump would be much higher than that actually observed if this additional depolarization did not induce an increase in the outward I_i (on account of I_K). The increase in repolarizing ionic current thus damps the retrograde depolarization wave which, as we now know, is the main cause of rhythm transformation in the nonuniform regions of the fiber.

We must conclude this section by emphasizing that compensatory changes in ionic currents are observed not only in the whole fiber but also in the uniform excitable membrane exposed to various factors modifying the conditions of generation of the membrane action potential.

The compensatory increase in I_{Na} (in its peak value and in the value of I_{Na} at the peak of the spike) explains the high stability of amplitude of the membrane action potential to a two- to fourfold decrease in maximum sodium permeability $\overline{P}_{Na}$ (Figs. 31 and 32) or in h_∞ (Fig. 33) of the node of Ranvier.

Compensatory changes in the amplitude of the second I_{Na} peak considerably weaken the effect of changes in reactivity of the potassium system (an increase or decrease in φ_K) on the duration of the descending phase of the nodal action potential (Fig. 48a,b), and so on.

Clearly, therefore, the parameters of both the membrane and propagating action potentials are reliably protected against external influences by positive and negative feedbacks in the membrane potential generators. Any deviation of these parameters from the initial values leads to a compensatory change in the ionic currents, aimed at reducing this deviation. That is why significant disturbances of generation and conduction of the action potential arise only if very large changes take place in the functional properties of the membrane or in the geometry of the fiber.

Chapter IX

Molecular Mechanisms of Ionic Permeability
of the Excitable Membrane

Despite the detailed quantitative description of the electrical behavior of excitable membranes, the physical mechanism of the ion movement through these membranes has still received relatively little study.

The many hypotheses which have been expressed in this problem can be divided into three main groups postulating the existence in the membrane 1) of mobile carriers, with affinity for particular ions; 2) of special pores, characterized by high ionic selectivity and the ability to open and close in response to changes of potential; and 3) of conformational changes in the membrane macromolecules in response to shifts of potential.

However, before turning to the examination of these hypotheses, it will be worthwhile considering, even if only briefly, the information given in the recent literature on the functional architecture of the membrane of nerve and muscle fibers.

Functional Architecture of the

Excitable Membrane

The view which is most widely accepted at the present time is that put forward by Danielli (see Davson and Danielli, 1943) and subsequently developed by Robertson (1964). According to this view, all cell membranes, whether excitable or inexcitable, have the same structure: they are formed from a bimolecular layer of liquids (phospholipids, cerebrosides, glycolipids, cholesterol) and

257

from two nonlipid monolayers covering it: a protein layer inside and protein–mucopolysaccharide layer outside (for a criticism of this hypothesis, see Korn, 1966, and the survey by Wolman, 1970).

The structural asymmetry of the membrane is matched by its functional asymmetry. This is demonstrated particularly clearly by comparison of the effects of lipolytic and proteolytic enzymes, various cations and anions, and pharmacological agents on the outside and inside of the membrane.

Enzymes

Investigations have shown that both the extracellular and intracellular action of phospholipases A and C on crustacean nerve fibers (Tobias, 1955, 1958, 1960; Narahashi, 1964) and nodes of Ranvier of frog nerve fibers (Nelson, 1958) cause depolarization of the membrane, a decrease in its capacitance and resistance, and consequently, a block to conduction. Electron-microscopic investigations have demonstrated considerable structural changes in cell membranes under these circumstances.

Since the target of these enzymes is lecithin, the structural and morphological disturbances found by the workers cited above have naturally been associated with the breakdown of this phospholipid. This conclusion is in good agreement with the fact that large quantities of lecithin are present in cell membrane (Brante, 1949).

By contrast with the phospholipases, proteolytic enzymes (trypsin, α-chymotrypsin, etc.) and also hyaluronidase (an enzyme splitting the mucopolysaccharide hyaluronic acid), when applied to the outer surface of nerve fibers, although producing definite structural changes in the membrane, do not, however, affect action potential generation (Tobias, 1955, 1958).

The proteases, however, are very effective if introduced inside the axon. Under these conditions, trypsin, chymotrypsin, and papain produce a considerable decrease in g_{Na} and g_K and block action potential generation (Rojas and Luxoro, 1963; Rojas, 1965; Tasaki and Takenaka, 1963; Takenaka and Yamagishi, 1969), while pronase (Rojas and Atwater, 1967) and Bacillus protease, prozyme, and bromelin (Takenaka and Yamagishi, 1969) produce a great prolongation of the action potential with the formation of a plateau

lasting several hundreds of milliseconds. These changes are irreversible, despite the fact that the resting potential and excitability of the axon remain unchanged during perfusion for several
hours. Electron-microscopic examination showed that the cytoplasm in these experiments had been almost completely removed.

On the basis of these results it was postulated that integrity
of the phospholipid layer in nerve fibers is important for maintenance of the properties of the resting membrane, while the protein structures of its inner surface enable this membrane to pass
into a state of excitation (see pp. 293-294).

Cations

It is well known that divalent cations, acting externally, have
a stabilizing and protective effect on the excitable membrane of
nerve and muscle fibers (see page 118).

By contrast, if Ba^{++}, Ca^{++}, Sr^{++}, or Mg^{++} ions are introduced
inside a giant axon, action potential generation is disturbed. For
instance, if the concentration of Ca^{++} in the internal solution is
20 mM, a block to conduction develops in approximately 5 min.
The same effect is produced by Mg^{++} ions in a concentration of
100 mM (Tasaki, Watanabe, and Takenaka, 1962).

Monovalent cations also differ in their effects on the outside
and inside of the membrane. For instance, Li^+ and Na^+ are the
most "favorable" for the outside, but the most harmful when injected inside the axon. In their ability to maintain excitability
inside a perfused axon, cations can be arranged in the following
order:

$$Cs^+ > Rb^+ > K^+ > Na^+ > Li^+,$$

where Li^+ is the most harmful.

It must be emphasized that the harmful action of monovalent
cations is much less marked than that of divalent cations (Tasaki,
Singer, and Takenaka, 1965).

Anions

The outside of the membrane of nerve fibers is relatively in-
sensitive to replacement of Cl^- by other anions (Lüttgau, 1960;
Hashimura and Osa, 1963; Koppenhöfer, 1964).

By contrast, a change in the anionic composition of the internal solution has a very marked effect on the electrical properties of the axon. Tasaki, Singer, and Takenaka (1965) compared the action of 15 anions on the inside of the membrane in two ways: by determining the length of survival of axons perfused with a solution containing the particular anion (with the cation unchanged), and by replacing one anion in the perfusion fluid by another anion. If, following such a replacement, the conduction block which developed in the axon under the influence of the first ion was removed, the second anion was regarded as "more favorable."

As a result of that investigation the following order was obtained: $F^- >$ $HPO_4^{--} >$ citrate $>$ tartrate $>$ propionate $>$ acetate $>$ $NO_3^- >$ $Br^- >$ $I^- \geq SCN^-$, in which F^- is the most, and I^- and SCN^- the least "favorable."

During perfusion of an axon with KI solution, a conduction block developed after 6 min. If, however, the axon was perfused with KF solution or, better still, with CsF solution, it remained capable of conducting high-amplitude action potentials for more than 5 h, despite the fact that the experiments were conducted at 21-23°C.

Since some of the most "favorable" anions have the property of binding divalent cations, the suggestion was made that the restorative effect of these anions is due to a decrease in the concentration of ionized calcium inside the axon. However, no simple correlation has yet been found between the equilibrium constant of the corresponding calcium salt and the physiological action of the anion. It has also been found that EDTA (ethylenediaminetetraacetate), which has much greater ability to bind Ca^{++} than any other anion, does not restore action potential generation in an axon perfused by an "unfavorable" anion. Tasaki and co-workers carried out several control experiments which showed that the "favorable" action of F^- ions is not due to their inhibitory effect on glycolysis.

Meanwhile it was shown that the order in which the inhibitory action of anions on the giant axon increases agrees well with the order of these ions in Hofmeister's lyotropic series: the greater the affinity of the anion for the positively charged groups of a protein and the weaker its salting-out effect in a colloidal solution, the stronger its inhibitory action (see Kruyt, 1949).

A similar rule was found in the series of cations given above: in this case, the inhibitory action was found to be closely connected with the affinity of the particular cation for the negatively charged groups of the phosphate colloid (the order of the cations was different for sulfate and carboxyl colloids). Tasaki and co-workers accordingly put forward the well justified hypothesis that both the anionic and cationic sequences found in the squid giant axon are directly connected with the influence of these ions on the positively and negatively charged groups of proteins and phospholipids.

Tasaki, Singer, and Takenaka (1965) regard the axon membrane as a macromolecular structure in which molecules of protein and phospholipids are linked together into complexes by salt bonds. The high membrane resistance in the resting state, in their opinion, is due to the dense packing of the macromolecules and the inaccessibility of most of the charged groups for transmembrane ionic transport. Addition of salts to the solution in adequate concentrations leads to the rupture or weakening of the bonds between the macromolecules, and thus increases the density of the ion-exchanging groups in the membrane. This increases its permeability (reduces its resistance) and depresses excitability. If the cationic or anionic affinity for fixed charged groups of the macromolecules is high, the tendency toward rupture of the bonds between the macromolecules of the membrane also becomes high, so that survival of the axon is shortened and the restorative property of the ion is reduced.

Tasaki attributes the functional asymmetry of the membrane to the fact that its outside, unlike its inside, is formed of a "triple complex" consisting of protein, phospholipid, and a divalent cation binding them together. The carboxyl group of the side chain may act in this complex like a colloidal anion. Since the affinity of these carboxyl groups for the alkali metals rises in the order $Li^+ < Na^+ < K^+$, potassium salts must be more effective at rupturing the complex than Na^+ and Li^+. In Tasaki's opinion, this order corresponds precisely to the order in which cations can suppress excitability when applied externally.

The different order of the cations, when applied intracellularly, is due to the fact that the inside of the membrane has a somewhat different structure from the outside and, in particular, the fixed negative charges (Teorell, 1964) on this side are formed, not of carboxyl, but mainly of phosphate groups of the biocolloid.

Dilution of Electrolytes in the Internal Medium

The view that fixed negative charges are present on the outside and inside of the membrane has received support from other investigators who have conducted experiments with internal perfusion of giant axons.

These investigations have shown that if the internal KCl solution is diluted with isotonic sucrose solution, action potential generation in the squid giant axon remains intact despite the very considerable decrease, or even total disappearance, of the resting potential. Under these circumstances the inactivation (h_∞ versus V) and sodium conductance (g_{Na} versus V) curves are displaced along the voltage axis toward higher values of the inside potential. This effect is connected with the decrease in ionic strength of the solution and not with the decrease in K^+ ion concentration in it, because it cannot be obtained by replacing K^+ by Li^+, Na^+, or Cs^+ ions or by choline.

If such a replacement is made, the decrease in resting potential is accompanied by disappearance of the action potentials (Narahashi, 1963; Baker, Hodgkin, and Meves, 1964; Moore, Narahashi, and Ulbricht, 1964).

To explain these facts, Chandler, Hodgkin, and Meves (1965) put forward the following hypothesis. Phosphate and other negatively charged groups are present on both sides of the membrane, but those on the outside are neutralized by Ca^{++} and Mg^{++} ions. Since ions can pass through the membrane, the fixed charges cannot make any direct contribution to the magnitude of the resting potential. The magnitude of the electrokinetic potential, i.e., the potential difference between the internal solution actually by the membrane and at a point some distance from it, must depend on the ionic strength of the solution and the density of the fixed charges. If the KCl solution is diluted from 300 to 6 mM the radius of the ionic atmosphere increases, and as a result the electrokinetic potential increases. Calculations showed that if the density of the charges was $1.4 \times 10^{13}/cm^2$ the electrokinetic potential on the inside of the membrane must change from -17 to -80 mV. Since the fixed negative charges on the outside of the membrane are

neutralized by divalent ions, the transmembrane potential difference for this dilution of the solution is 80 mV, despite the fact
that there may be no potential difference between the external and
internal solutions in this case. In these workers' opinion, these
findings explain the changes in threshold and in the inactivation
curve of the axon observed following dilution of the internal solution, for the sodium permeability must depend to a far greater
extent on the potential on the membrane than on the complete potential difference between the internal and external solutions.

According to this hypothesis, changes in the resting potential,
critical level of depolarization, and level of the peak of the spike
on dilution of the internal KCl solution by nonelectrolytes are therefore purely apparent and are attributable to inability to measure
the true value of the transmembrane potential difference.

However, facts have been obtained which contradict this interpretation of the dilution effects. For instance, in the work of Tasaki,
Singer, and Takenaka (1965) cited above it was found that during
internal perfusion of the giant axon with a 500-600 mM KCl solution the electrical properties of the fiber remain unchanged only
for the first 10-20 min, after which the resting potential, action
potential, and resistance of the membrane start to fall. After 30-
40 min conduction of impulses ceased completely in the perfused
segment of the axon. However, it can be restored at once if, after
development of the block (but not later than 1 min after its development) the 500 mM KCl solution is replaced by 200 mM KCl
(isotonicity is maintained with sucrose or glycerol). The restorative effect is more marked still on further dilution of the solution.
It is interesting that restoration of action potential generation in
all these cases takes place despite a further decrease in the transmembrane resting potential (E_m), due to a decrease in the concentration gradient for K^+ ions.

It follows from all these results that a decrease in the ionic
strength of the internal solution has a considerable effect on the
physicochemical state of the excitable membrane. The possibility cannot therefore be ruled out that the preservation of its
ability to generate high-amplitude action potentials despite the
sharply reduced resting potential is not apparent, but a true phenomenon associated with actual weakening of the original sodium
inactivation.

In experiments with internal perfusion of giant axons the following very important fact also was discovered. Chandler and Meves (1966, 1970) and Adelman and Senft (1966) showed that K^+ ions exert a specific effect on the inactivation process. With the complete removal of K^+ from the internal solution (replacement of 300 mM KF by 300 mM NaF) inactivation of sodium permeability becomes incomplete, so that during prolonged fixed depolarization of the membrane the sodium current can continue for several tens of seconds longer. Correspondingly, the action potential of such an axon acquires a very long plateau. The most interesting fact is that under the conditions being considered, the decrease in h_∞ accompanying changes in V toward depolarization is replaced by a slight increase when V begins to approach the level of the sodium equilibrium potential V_{Na}.*

It might be supposed that these effects are due to disappearance of the outward potassium current. However, experiments showed that the inactivation under these conditions is also dependent on the external K^+ ion concentration: the lower this concentration the less marked the inactivation and the long plateau of the action potential (Adelman, Dyro, and Senft, 1965a,b; Adelman and Palti, 1969).

Poisons

It was stated above that tetrodotoxin specifically blocks the initial sodium ionic current while leaving the delayed potassium current virtually unchanged (see page 73). Special investigations have shown that this effect of tetrodotoxin can be obtained only if the poison is applied externally to the axon. Injection of tetrodotoxin into the giant fiber, even in concentrations a hundred times greater than those effective on external application, produces no change in the sodium current (Narahashi, Anderson, and Moore, 1966). Similar results were obtained with medullated nerve fibers when tetrodotoxin was applied to the inside of the membrane of a node of Ranvier through the cut neighboring internode (Koppenhöfer and Vogel, 1969).

These results show that the receptor groups with which tetrodotoxin interacts lie on the outside of the membrane and that

---

* Similar changes in the h_∞ versus V curve were observed by Koppenhöfer and Schmidt (1968) when a node of Ranvier was treated with scorpion venom.

tetrodotoxin is unable to diffuse through the membrane from the internal into the external solution, as procaine does (Truant, 1950). This conclusion accords well with the insolubility of tetrodotoxin in organic solvents (Mosher et al., 1964).

Unlike tetrodotoxin, tetraethylammonium (an agent specifically blocking the potassium current) is effective only when introduced inside the squid giant axon (Tasaki and Hagiwara, 1957). A similar action on the potassium current, limited to the inside of the membrane, is produced in the giant axon by Cs^+ ions (Chandler and Meves, 1965; Adelman and Senft, 1966; Sjodin, 1966). This indicates that the areas with which tetraethylammonium and Cs^+ ions interact lie on the inside of the membrane in this object.

However, in most other excitable structures tetraethylammonium is effective on external application also: this is true in the case of frog medullated nerve fibers (Tasaki, 1959a; Lüttgau, 1960; Khodorov and Belyaev, 1965a,b; Schmidt and Stämpfli, 1966; Koppenhöfer, 1966, 1967; Hille, 1967a,b), crab nerve fibers (Burke, Katz, and Machne, 1953), neurons of Ochnidium verruculatum (Hagiwara and Saito, 1955), frog skeletal muscle fibers (Hagiwara and Watanabe, 1955), and the Purkinje fibers of warm-blooded animals (Halidman, 1963).

To understand the mechanism of action of tetraethylammonium (TEA), it is essential to remember that its injection inside the giant axon (40 mM) almost completely blocks the outward I_K while producing virtually no change in the inward I_K (arising in response to hyperpolarization of the membrane after its preliminary depolarization). In low concentrations of TEA, a normal I_K arises initially in response to a change in V toward depolarization, but it begins to fall after a fraction of a millisecond (Armstrong, 1966). This delay in its decrease is particularly marked if, instead of TEA, the propyl-triethylammonium ion is applied. In this case the temporal course of the changes in I_K on depolarization becomes qualitatively very similar to that characteristic of I_{Na} (Amstrong, 1968). These results led this investigator to conclude that the TEA ion occupies a blocking position in the potassium channel only after it has been opened by a depolarizing stimulus. Moreover, TEA ions enter these channels only if the ionic current is outward in direction.

By contrast with TEA and tetrodotoxin, with their different actions when applied on the outside and inside of the membrane, procaine is effective whether applied externally or internally to the giant axon membrane (Narahashi, Anderson, and Moore, 1967). This is evidently because the changes in ionic permeability produced by the action of procaine (like other local anesthetics) are associated with their penetration into the lipid layer of the membrane.

These results indicate that the excitable membrane is a highly specialized structure, the inner and outer surfaces of which differ in their properties and functions.

Let us now turn to the examination of some hypotheses concerning the mechanisms of passive ion movement through this membrane.

The Hypothesis of Ionic Carriers

In their investigation of ionic currents in squid giant axons Hodgkin, Huxley, and Katz (1949) found an exponential relationship between the peak g_{Na} and the initial potential shift. A change in E toward depolarization by 4-6 mV corresponded to an e-fold increase in g_{Na}. To explain the steepness of the g_{Na} versus E curve these workers initially postulated that changes in g_{Na} are associated with displacement of mobile carriers inside the membrane under the influence of the electric field. These carriers may be lipid-soluble particles with affinity for Na^+ or K^+ ions. According to this hypothesis, Na^+ crosses the membrane not as a free ion, but in combination with the lipid-soluble carrier, which has a high negative charge but which cannot bind more than one Na^+ ion.

Since the carrier remains negatively charged even after binding the Na^+ ion, in the resting fiber it is attracted to the outside of the membrane. Depolarization of the membrane enables movement of the carrier, together with a Na^+ ion to the inside of the membrane, where the Na^+ ion detaches itself from the carrier and enters the cytoplasm.

According to this hypothesis it would be expected that during depolarization of the membrane an outward current would be generated first, because the negatively charged particles of the carrier moved from the outside to the inside of the membrane. This outward current must be replaced by an inward current only when

the carrier gives up the sodium ion to the internal solution and itself returns to the outside of the membrane.

In reality, as voltage-clamp experiments showed, the inward
sodium current is never preceded by outward current. An outward current arises only during longer depolarization and is connected with the transport of K^+ ions.

In their later communications (Hodgkin and Huxley, 1952d;
Hodgkin and Keynes, 1955b, Hodgkin, 1964) these workers accordingly rejected the carrier hypothesis (a role was ascribed to carriers only in active transport), and it was postulated that Na^+ and K^+
ions diffuse through special "pores" — channels whose permeability
to these ions is regulated by certain charged particles with the
ability of moving inside the membrane in response to changes in
potential (see page 85).

However, in recent years interest in the question of ionic carriers has revived in connection with results obtained by the study
of artificial phospholipid membranes during the action of macro-
cyclic antibiotics (see page 293).

Second, new ways of testing the validity of this hypothesis
have been developed. Theoretical studies by Eisenman, Sandlom,
and Walker (1967) have shown that if the ionic current is conducted
through a membrane by carriers (by "mobile ion-exchange sites")
there must be a limit beyond which this current cannot rise with
an increase in the transmembrane potential difference (the saturation phenomenon). However, the experiments of Moore (1967) on
single nodes of Ranvier, by the voltage-clamp method, showed that
when potassium peremability is at its maximum the potassium current does not exhibit saturation even when the potential on the inside of the membrane is increased from −70 to +100 mV or more.
Throughout this range of variation of potential, the momentary values of I_K increase linearly. In Moore's opinion, these results rule
out any possible role of carriers in the conduction of the ionic current through the membrane.

The Hypothesis of Ion Pores

The suggestion that cell membranes contain special "pores"
with selective permeability relative to particular ions is the most

widely accepted hypothesis in contemporary general physiology. This hypothesis was put forward on the basis of studies of the mechanisms of ion diffusion through artificial precipitation and collodion membranes, and it was subsequently confirmed by experimental results obtained with membranes of erythrocytes and nerve and muscle fibers.

Experiments on these objects showed that cell membranes are permeable not only to water and ions, but also to many small neutral molecules, soluble in water but not in lipids.

The most conclusive results from this standpoint were obtained by Solomon and his collaborators (Goldstein and Solomon, 1960; Whittembury, Sugino, and Solomon, 1960), conducted on membranes of erythrocytes, the squid giant axon, skeletal muscle, and kidney tissue.

The effective radius of the membrane pores determined by these workers on the basis of osmotic measurements was found to be about the same for a wide class of neutral molecules.

The permeability of the squid giant axon membrane for neutral molecules has more recently been investigated by Villegas et al. (1966), who used for this purpose C^{14}-labeled erythritol, mannitol, and sucrose.

Other evidence of the existence of pores in the membrane was obtained by Stallworthy and Fensom (1966), who found that when ions passed through the squid giant axon membrane under the influence of an applied external field, electron-osmotic transport of water occurs simultaneously: for each ion passing through the membrane, there are about 26 molecules of water.

However, to explain all the features of passive ion movement through a membrane by the "pore" hypothesis it has to be accepted that besides nonspecific "leak current" pores there are also ion-selective sodium and potassium pores, and that these pores are equipped with a special mechanism (known as "gates") which enable the pores to be opened and closed in response to changes in membrane potential.

The number of sodium channels has been estimated from the binding of tetrodotoxin (see Moore et al., 1967). Calculations show that a single sodium channel can pass Na^+ ions at rates over 10^8

sec^{-1}, which is 10^6 times higher than the turnover rate of the sodium pump (Hille, 1971). Recent published material on these matters follows.

Specificity of the Ion Channels

It was stated above that the sodium and potassium currents in nerve fibers are relatively independent and that chemical compounds which specifically block the increase in sodium (tetrodotoxin) and potassium (tetraethylammonium) permeability of the membrane during depolarization exist.

These findings suggest that the channels through which Na^+ and K^+ ions move are chemically distinct and spatially separate (Hille, 1967a, 1970).

The results obtained by studies of the action of the insecticide DDT on nerve fibers provide convincing evidence, according to Narahashi and Moore (1968) and Hille (1970), in support of this view.

In the squid giant axon, under the influence of DDT the process of blocking the sodium channels (a decrease in m) takes place much (about 4.5 times) more slowly, whereas the activation of potassium permeability (an increase in n) is only slightly (1.4-1.6 times) retarded. In the node of Ranvier these differences in the temporal course of the changes in P_{Na} and P_K under the influence of DDT are more marked still (Hille, 1968). These facts can hardly be reconciled with the existence of a common path for Na and K ions in the membrane postulated by Mullins (1968), even if it is assumed that the sodium channel is converted into a potassium channel only after completion of its inactivation (see Hille, 1970, 12).

The selectivity of the channels relative to different ions, as already mentioned (see pp. 75-76), is not absolute. In the perfused squid axon, for instance, the sodium channels (or, more precisely, channels blocked by tetrodotoxin) are characterized by the following permeability ratios for monovalent cations (Chandler and Meves, 1965):

$$P_{Li} : P_{Na} : P_K : P_{Rb} : P_{Cs} = 1.1 : 1 + 1/12 : 1/40 : 1/61. *$$

---

*These results were confirmed by the work of Rojas and Atwater (1967a), Atwater, Bezanilla, and Rojas (1969), and Binstock and Lecar (1969).

The sodium channels are also permeable to ammonium and hydroxylammonium ions and to organic cations such as guanidinium and hydrazinium.

This explains the fact that the above-mentioned cations can replace Na^+ ions during action potential generation in nodes of Ranvier (Lorente de Nó, Vidal, and Larremendi, 1957; Lüttgau, 1958; Deck, 1958), spinal ganglion cells (Koketsu, Cerf, and Nishi, 1959), and squid giant axon (Tasaki, Singer, and Watanabe, 1965).

Penetration of guanidinium inside the perfused giant axon during action potential generation was demonstrated by direct experiments with C^{14}-labeled guanidine. The production of an inward current in an axon immersed in hydrazine solution was obtained in experiments conducted by the voltage-clamp techniques (Tasaki, Singer, and Watanabe, 1967). Action potentials generated in solutions of these organic cations are inhibited just as well by tetrodotoxin as sodium action potentials (Tasaki, Singer, and Watanabe, 1966).*

No quantitative assessment of the relative permeabilities of the sodium channel to different organic cations has yet been undertaken. However, the ability of ammonium cations to restore action potential generation in sodium-free solutions has been shown to decrease with an increase in the number and length of the carbon chains added to the nitrogen atom. Moreover, cations containing hydroxyl groups are much more effective than those with amino, hydrogen, alkyl, or phenyl groups (Tasaki, Singer, and Watanabe, 1965).

Binstock and Lecar (1967; 1969) found that the ratio between the permeabilities of the early tetrodotoxin-sensitive channel of the squid giant axon to Na^+ and NH_4^+ ions is approximately 3:1. In the node of Ranvier, permeability of the membrane to guanidini-

---

*A fact of great interest is that for an action potential to develop in a giant axon after total replacement of sodium ions by organic cations, the internal cation must be Cs^+ or Rb^+. Replacement of Na^+ by guanidinium or hydrazinium in intact axons has no effect (Tasaki, Singer, and Watanabe, 1965). We think it is explained by the fact that, with a normal outward I_K, the inward flux of organic cations is too weak to cause regenerative depolarization of the membrane. This can occur only if the outward current has been greatly reduced, and this can be achieved by replacing the internal K^+ by Rb^+ and Cs^+.

um ions is approximately 1/10 (Lüttgau, 1958) and for ammonium ions 1/5 (Dodge, 1963) the maximum sodium permeability.

Specificity of the potassium channels has received less study. In the squid giant axon the relative permeabilities of the potassium channel for K:Rb:Cs are approximately 1:0.8:0 (Pickard et al., 1964). The NH_4^+ ion passes more readily through the potassium channels than the sodium: the ratio between the permeabilities of the potassium channel to K^+ and NH_4^+ is 3:2 (Binstock and Lecar, 1967, 1969).

How can this selectivity of the ion channels be explained?

Investigations on artificial membranes (see Eisenman, Sandlom, and walker, 1967) have shown that if fixed negative charges are present in the walls of these channels their cationic permeability can be determined by the following principal factors: 1) the available pore area of the membrane; the channels; 2) the geometry of the channels (the shape of the "entrance," the tortuosity of the pore, and so on); 3) electrostatic interaction between the cations and anionic sites in the walls of the pores ("ion−site" and "ion−matrix" interaction); 4) solvent drag phenomena (electroosmosis).

If the pore diameter is large enough for ions to penetrate together with the solvent, the permeability of the membrane for these ions will be determined primarily by the 1st, 2nd, and 4th of these factors. In this case mobility of the ions in the membrane channels and the energy of their activation will be of the same magnitude as when these ions are in aqueous solution. Processes of ion exchange with the anionic sites of the channels walls have no significant effect on their permeability.

If, however, the pore diameter is small, the importance of the 3rd and 4th factors becomes dominant, and in this case the properties of the channels come to resemble those of an ion exchanger for which specific differences between cations carrying an equal charge (for example, Na^+ and K^+) become very significant. Under these circumstances the mobility of the cations is reduced while the activation energy, on the other hand, is increased.

However, a clear understanding of these qualitative differences between the properties of "wide" and "narrow" pores has been obtained only recently, with the development of the theory of ion-selective glass membrane (see below). Previously, for several years attempts have been made and, indeed, they are still being made at the present time, to explain the selective ionic permeability of excitable membranes purely on account of inequality of the "hydrated" radii of penetrating and nonpenetrating cations. The increase and decrease in ionic permeability in response to changes of potential were regarded purely as the result of widening and narrowing of the ion pores (Shanes, 1958; Mullins, 1960; Liberman, 1963).

The speculative character of these hypotheses became evident after the sodium and potassium channels had been identified with the aid of tetrodotoxin and tetraethylammonium and the relationships between their permeabilities for different inorganic and organic cations, which we have already examined above, had been established.

The radius of the hydrated ion is usually estimated from its mobility in an electric field. This mobility diminishes in the order: $Cs^+ > Rb^+ > K^+ > Na^+ > Li$.

The ionic permeability of the membrane ought to diminish in the same order, because lower mobility of an ion in solution corresponds to a larger radius of the ion in the hydrated state.

However, behavior of this type has been found only in artificial collodion membranes working on the "molecular sieve" principle (Sollner, 1955; Dray and Sollner, 1956). The ion channels of the excitable membrane of the nerve fiber are characterized, as we have seen, by a totally different order of permeabilities, for Cs^+ ions pass with great difficulty both through the sodium and, in particular, through the potassium channels.

To explain this behavior of the ions, so paradoxical from the point of view of the molecular sieve theory, Mullins (1964) suggested that the generally accepted method of estimating the radius of a hydrated ion from its mobility in an electric field is inaccurate and that, in fact, this radius is computed from the radius of the ion in the crystal lattice and the diameter of a molecule of water (Fig. 97). These "monohydrated" ions can penetrate through the cell membrane, in Mullins' opinion, only if the radius corresponds sufficiently closely to the radius of the pore. In that case the walls of the pore act as replacements for the remaining hydra-

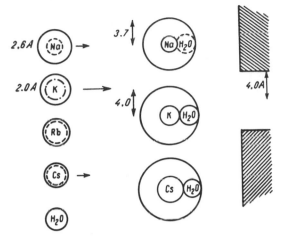

Fig. 97. Comparison of the radii of hydrated ions calculated from values for permeability and the radii of
ions surrounded by a single layer of water molecules.
Usual radii of hydrated ions and dimensions of the water
molecule are shown on the left. The broken circles
correspond to "crystal radii." Arrows show relative
permeability of muscle fiber membrane to these ions
(Mullins, 1964).

tion membranes, and this replacement takes place without any
change in the free energy of the ion. If, however, the radius of the
monohydrated ion is substantially greater or less than the radius
of the pore, the free energy of hydration, holding the ion in solution, prevents its passage into the membrane.

The radii of the monohydrated ions calculated by Mullins'
method lie in the same order relative to one another as the order
discovered by Chandler and Meves for the sodium channel. However, it is not clear how this explains the fact that organic cations
such as hydrazinium and guanidinium, whose crystal radii (2.2 and
2.7 Å) are respectively more than twice the crystal radius of the
sodium ion (0.97 Å; Hille, 1967b), can also pass through the sodium
channel. If it is assumed that ions of hydrazine and guanidine pass
through the membrane in a state of monohydration, their radii must
be taken as 4.93 and 5.43 Å, respectively, which is much greater
than the radius of the monohydrated sodium ions (3.7 Å).

Attempts to link selective ionic permeability of the membrane
with the character of interaction between cations and the fixed

charges of the walls of the corresponding channels are much more promising.

Chandler and Meves (1965) stressed the great resemblance between the order of the cations obtained for the sodium channel and that found for sodium-selective glass electrodes. Nikol'-skii (1937) and Eisenman (1964) have developed a neat theory of the ionic selectivity of glass membranes. These workers showed that negatively charged fixed loci, performing ion-exchange functions, exist in such membranes. The selectivity of glass relative to particular ions is determined by the ratio between the free energy of interaction of the given cation with molecules of water, on the one hand, and with the anionic loci of the glass on the other. The latter apparently compete with water for interaction with the cations. Clearly, the higher the energy of electrostatic interaction between the cation and negatively charged locus of the glass, the easier it will be for the cation to pass from solution into the glass and, conversely, if the energy of interaction with the glass is weak, the ions will be held in solution by the water molecules.

The selectivity of different glasses depends on the nature of the anionic locus. There are some loci (of the $AlOSi^-$ type) for which the electrostatic energy of interaction with cations is below their free energy of hydration. Such loci characteristically are arranged in the order: $Cs^+ > Rb^+ > K^+ > Na^+ > Li^+$.

This order is determined by the ease with which the cation is removed from the water membrane. The hydration energy of the Cs^+ ion is the smallest, so that it passes from aqueous solution into the glass more easily than the rest.

The order with which the cations penetrate into glass from areas in which the electrostatic energy of interaction with cations is much greater than the free hydration energy (loci of the SiO^- type) is diametrically opposite: $Li^+ > Na^+ > K^+ > Rb^+ > Cs^+$.

In this case the greater the force with which the locus attracts a given cation, the greater the probability that this cation will migrate from the aqueous solution into the glass. This force is inversely proportional, however, to the crystal radius of the cation.

The loci in glass are formed by oxygen-containing anions located on the surface of the inorganic polymer. From this point of view glass is similar to the biological polymers in the cell membrane or cytoplasm, containing phosphate and carboxyl groups. For example, a locus of SiO^- type, characterized by an order of selectivity $Li^+ > Na^+ > K^+ > Rb^+ > Cs^+$, is geometrically very similar to the $H_2PO_4^-$ anion or to the phosphate locus in phosphoproteins, phospholipids, nucleic acids, or the ATP molecule. In all these cases, the nearest neighbor of the cation penetrating into a membrane from solution is oxygen, the effective field of which is modified by the more distant C, P, Al, or Si atoms.

Calculations have shown that with variations in the strength of the effective anionic field, besides the orders of anions given above, a further nine other orders, occupying an intermediate position, may exist in different glasses. They include the order of selectivity characteristic of the resting membranes of nerve and muscle fibers: $K^+ > Rb^+ > Cs^+ > Na^+ > Li^+$ (Conway and Moore, 1954; Mullins, 1959; Sjodin, 1959).

While mentioning the possible similarity between the mechanisms of ion movement in a sodium-selective glass electrode and in the active cell membrane, Chandler and Meves (1965) also emphasized some significant differences between them. For instance, the dry sodium-selective glass (NAS 11-18) which they used, in a thickness of 50 Å, has a resistance of the order of 1400 Ω/cm^2, whereas the resistance of the active membrane of approximately the same size is only 5-30 Ω/cm^2.

To understand the principle underlying this difference, it must be assumed that either ion-dipole interaction takes place in cell membranes between ions and the structural elements of the membrane (hence the high dielectric constant of the membrane), or the ions cross the membrane in a partly hydrated form.

The second of these factors appears very important, because it explains the ability of Na^+ and guanidinium to replace each other in action potential generation.

Paoloni (1963) pointed out the great resemblance between the electronic structure of guanidinium and the hydrated sodium ion

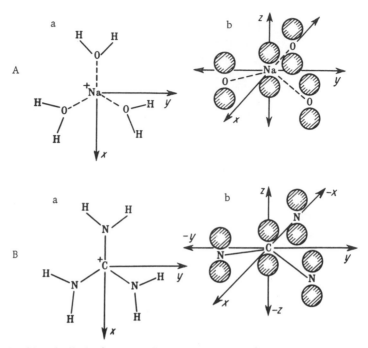

Fig. 98. Similarity between electronic structure of the hydrated sodium ion $Na(H_2O)_3^+$ (above) and the guanidinium ion $C(NH_2)_3^+$ (below). A, a) Arrangement of coordination bonds; b) arrangement of coordination orbitals; Na orbital is not filled, but each orbital of the oxygen atoms is filled by two electrons. B, a) Arrangement of bonds; b) arrangement of orbitals; carbon orbital is not filled, but each orbital of the nitrogen atoms is filled by two electrons (Paoloni, 1963).

(Fig. 98). He accordingly postulated that for a given ion to pass through a membrane, not only the size of the ion, but also the character of its field of force, is of decisive importance. The differences in the electronic structure of Na^+ and K^+ ions indicate that the sodium and potassium channels differ not only in size, but also in the physicochemical structure of their inner surface.

The effects of tetrodotoxin receive a temptingly simple explanation from this point of view. Tetrodotoxin incorporates a guanidinium group (Fig. 99), and it can be accordingly be postulated that during the action of tetrodotoxin the guanidinium group enters a sodium channel, while the rest of the molecule prevents the passage of the toxin through this channel, thus effectively blocking it.

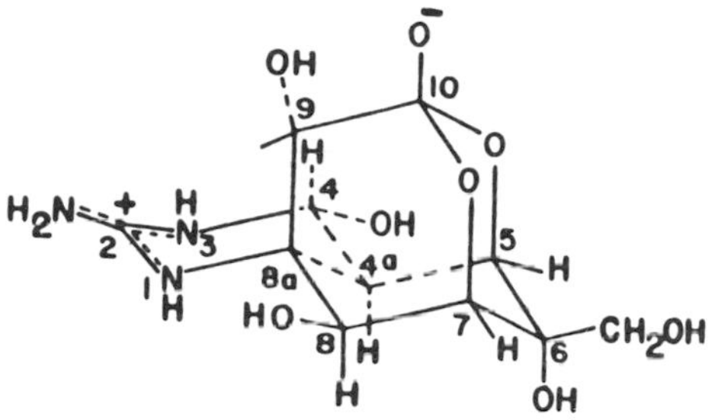

Fig. 99. The structure of tetrodotoxin (Kao, 1966).

In this explanation the reason for the high activity and specificity of action of the poison may be that the structures of the membrane around the sodium channel are complementary to those parts of the tetrodotoxin molecule which lie stereochemically close to the guanidinium group. A similar mechanism of "blocking" of the sodium channels, according to the hypothesis just considered, underlies the action of another powerful poison — saxitoxin, which differs from tetrodotoxin in its structure but which also contains a guanidinium group (see Kao and Nishijama, 1965; Kao, 1967).*

If this hypothesis is confirmed, there is hope that a closer insight will be obtained, with the aid of these toxins or their analogs, into the molecular architecture of the gates of the sodium channels.

One difficulty with which Paoloni's hypothesis has to contend, however, is that the sodium channels of an active membrane are accessible not only to guanidinium, but also, has been pointed out, to other organic cations such as hydrazinium ($NH_2-NH_3^+$), which differ completely from the hydrated sodium ion in their electronic structure.

The Possible Role of Ca^{++} Ions in the Mechanism of

Opening and Closing of the Ionic Channels

So far we have examined only the factors which determine the specificity of the sodium and potassium channels of a membrane

---

*Further light on the mechanism of action of tetrodotoxin is shed by the results published by Narahashi, Moore, and Poston (1967b) and obtained during the investigation of the effects of derivatives of this compound.

Let us now turn to the possible mechanisms of changes in their ionic permeability on excitation.

Most of the hypotheses relating to this question are developments of the idea of Gordon and Welsh (1948) that Ca^{++} ions play a leading role in this process. It is well known that a decrease in the $[Ca]_0$ concentration leads to an increase in excitability and often to the generation of self-sustained rhythmic ("automatic") activity;[*] with an increase in $[Ca]_0$, on the other hand, the excitability of nerve and muscle fibers is sharply reduced (see Chapter I). In this respect the effects of a decrease and increase in $[Ca]_0$ are similar to the effects of depolarization and hyperpolarization of the membrane (see the surveys by Brink, 1954; and Shanes, 1958).

The quantitative aspect of these relationships has been studied in detail by Frankenhaeuser and Hodgkin (1957) in experiments by the voltage-clamp method on squid giant axons.

Their work showed that the degree of increase in sodium conductance of the membrane on depolarization depends on $[Ca]_0$. The higher this concentration, the smaller the increase in g_{Na} for a given depolarization of the membrane. It is only the maximum to which the sodium conductance rises during very strong depolarization that is independent of $[Ca]_0$. The whole curve of g_{Na} as a function of membrane potential (g_{Na} versus V) is therefore shifted toward higher values of V. The effect of Ca^{++} ions on the reactivity of potassium conductance and of the system of sodium inactivation is similar (for details, see pages 124 and 128).

To explain these facts, Frankenhaeuser and Hodgkin postulated that Ca^{++} ions play the role of particles blocking the ionic channels in the membrane. At rest, i.e., when the transmembrane potential difference is high, Ca^{++} ions are held in these channels by forces of the electrical field. During depolarization of the membrane or with an increase in $[Ca]_0$, the time during which the channels are free from calcium increases, with the result that the sodium (and potassium) permeabilities increase. Opposite changes in g_{Na} and g_K take place during hyperpolarization of the membrane or an increase in $[Ca]_0$.

---

[*] The ionic mechanism of generation of this activity has been analyzed by Khodorov, Grilikhes, and Timin (1970).

Simple calculations showed, however, that if this hypothesis is true, with an e-fold increase in $[Ca]_0$ the shift of the g_{Na} versus V curve must be not less than 12.5 mV.* In reality, however, both in the squid giant axons (Frankenhaeuser and Hodgkin, 1957) and in nodes of Ranvier of frog nerve fibers (Hille, 1967a) it does not exceed 9 mV. In crustacean giant fibers it is lower still, about 6 mV (Blaustein and Goldman, 1966a).

Another defect of this hypothesis is that it does not explain the increase in calcium permeability of the membrane on depolarization. An increase in the inward calcium current during repetitive stimulation of the giant axon was observed by Hodgkin and Keynes (1957) in experiments with the isotope Ca^{45} (see also Baker, 1971). Under certain conditions this increase in calcium permeability becomes so great that Ca^{++} ions can carry enough inward current to generate an action potential.

Watanabe et al. (1967), for instance, observed the generation of long calcium action potentials in squid giant axons during internal perfusion with CsF solution. All monovalent cations were removed from the external solution and replaced by 200 mM $CaCl_2$. The osmotic pressure was maintained with glycerol. Since the action potentials arising under these conditions were completely suppressed by tetrodotoxin, it can be assumed that at a time of activity the Ca^{++} ions cross the membrane through the sodium channels (Fig. 100).

Finally, it is very difficult to understand, on the basis of Hodgkin's hypothesis, the fact that Ca^{++} ions do not compete with tetrodotoxin for the sodium channel. An increase in $[Ca]_0$ does not weaken the inhibitory action of tetrodotoxin on the membrane on a single node of Ranvier, as would be expected if such competition existed (Khodorov and Belyaev, 1967).

---

*If the ratio between the number of channels closed by Ca^{++}(P) and the number of open channels $(1 - P)$ is taken as proportional to $[Ca]_0$, then

$$P/(1 - P) = b[Ca]_0 \exp(-2\delta EF/RT)$$

where δ is the fraction of the potential moving Ca^{++} from the external solution to the point where this ion blocks the channel. If this point lies in the middle of the membrane $\delta = 0.5$, but if it is closer to the inside of the membrane $\delta \cong 1$; b is a constant. In the latter case an e-fold change in $[Ca]_0$ is equivalent to a change in E by 12.5 mV.

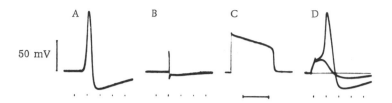

Fig. 100. Oscillograph records demonstrating development of an "all-or-nothing" action potential in an axon perfused internally with sodium phosphate solution and externally with $CaCl_2$ solution. Records A and B were taken before initiation of internal perfusion; the external medium contained 300 mM NaCl and 100 mM $CaCl_2$ in A and only 100 mM $CaCl_2$ in B. The time markers are 1 msec apart. Record C was taken approximately 12 min after the onset of internal perfusion with 10 mM sodium phosphate; the external medium was 100 mM $CaCl_2$ (as in B). The time marker represents 10 sec. Record D was obtained from the same axon after switching the internal perfusion fluid to 400 mM KF and the external medium to the solution used in A; sub- and suprathreshold responses are superposed. Time markers are 1 msec apart. The stimulus duration was 0.1 msec for A and B, 100 msec for C, and about 0.3 msec for D. Axon diameter: approximately 400 μ. 21°C (Watanabe, Tasaki, Singer, and Lerman, 1967).

For a general assessment of this hypothesis it is most essential to realize that most effects of Ca^{++} ions can be successfully explained by a completely different hypothesis, put forward by Huxley (cited by Frankenhaeuser and Hodgkin, 1957). Huxley postulated that Ca^{++} ions are adsorbed on the outer surface of the membrane, because of the presence of fixed negative charges there. Binding of the Ca^{++} ions by these charges leads to the generation of an electrical field in the membrane, which is added to the resting potential (although it cannot be detected by the usual methods of measurement of the potential difference between the external solution and the contents of the fiber). This hypothesis readily explains the linear relationship between the shift of the g_{Na}, g_K, and h_∞ curves along the V axis and the concentration of Ca^{++}, Ni^{++}, Co^{++}, Zn^{++}, La^{+++}, or H^+ ions in the medium (Frankenhaeuser and Hodgkin, 1957; Hille, 1968a; Blaustein and Goldman, 1966, 1968; Khodorov and Peganov, 1969; Mozhaeva and Naumov, 1970). The absence of competition between Ca^{++} ions and tetrodotoxin can also be understood: these agents evidently interact with different receptor groups of the membrane, and a change in the electric field in the membrane is known to have no effect on the value of $\overline{g}_{Na}$. Finally, the restorative action of an excess of Ca^{++} ions on procainized or depolarized nerve and muscle fibers (page 118) can be explained, from the point of view of this hypothesis, by weakening of inactivation of the sodium system as the result of the increase in potential on the membrane.

However, facts have been obtained which indicate that the effects of Ca^{++} ions cannot be reduced purely to their influence on the electrical field of the membrane. Let us examine some of them.

1. It was mentioned above that an increase in $[Ca]_0$ not only increased φ_{Na}, but also considerably accelerates the decrease in g_{Na} on removal of the depolarization at the moment when g_{Na} reaches its maximum. Conversely, a decrease in $[Ca]_0$ slows this process considerably. In connection with this problem it is essential to note that these kinetic effects are much stronger than those which would be expected from the rule that "an e-fold increase in $[Ca]_0$ is equivalent to a change in potential of 9 mV" (Frankenhaeuser and Hodgkin, 1957).

Nickel ions, with the same valence as Ca^{++}, have the directly opposite effect on the kinetics of ionic permeabilities of the membrane (page 119). In many respects their effect resembles that of cooling the fiber on τ_m, τ_n, and τ_h. These processes are retarded even more strongly by La^{+++} ions (Hille, 1968a; Blaustein and Goldman, 1968; Takata et al., 1966). Changes in the electrical field in the membrane cannot possibly explain all these findings.

2. A direct indication of a change in the molecular structure of the membrane under the influence of polyvalent cations is given by the following facts.

In experiments on single Ranvier nodes, Ooyama and Wright (1961) found that during very strong hyperpolarization of the membrane, there are signs of the development of "breakdown" of the membrane, with the appearance of irregular fluctuations of potential and, ultimately, a marked decrease in the potential, which is clearly linked with a decrease in resistance of the membrane. If the hyperpolarizing current is very strong, the breakdown of the membrane is total (Stämpfli and Willi, 1957), but with weaker hyperpolarization the breakdown phenomena are completely reversible.

Joint experiments with E. M. Peganov (Khodorov and Peganov, 1969) showed that the resistance of the membrane to this action of the hyperpolarizing current depends on the Ca^{++} ion concentration in the solution: the higher the value of $[Ca]_0$, the greater the transmembrane potential difference at which the signs of breakdown appear.

This relationship was observed particularly clearly when the effect of Ca^{++} ions on the generation of "hyperpolarization responses" (HRs) was studied.

To obtain such responses (Segal, 1958; Stämpfli, 1959; Tasaki, 1959a; Mozhaeva, Mozhaev, and Skopicheva, 1966) the membrane of a nerve fiber is depolarized by increasing the K^+ ion concentration in the solution, and a hyperpolarizing direct current, of different strength, is then applied to the membrane. When this cur-

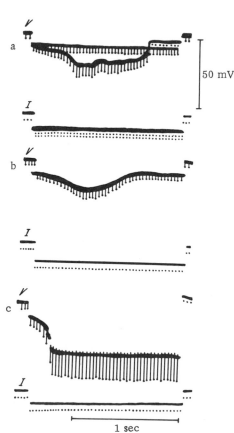

Fig. 101. Effect of Ca^{++} ions on hyperpolarization responses in a single node of Ranvier kept in isotonic KCl solution. a and b) Abortive hyperpolarization responses of a Ranvier node after a long stay in KCl solution containing traces (~ 0.01 mM) of Ca^{++} ions; c) restoration of hyperpolarization response after addition of 10 mM $CaCl_2$ to the medium. Top beam (V) shows changes in potential, bottom beam (I) current applied. Short (5 msec) pulses of current applied rhythmically to the node so that changes in resistance of the membrane can be studied (Khodorov and Peganov 1969).

50 mV

1 sec

rent reaches a certain threshold value, the change in potential
produced by the current assumes the character of a double step
(Fig. 101c). The appearance of the second step is due to an increase in resistance of the membrane which, in turn, is caused by
the regenerative decrease (restoration) of P_K to its initial low
values. The way in which the process develops is as follows: the
current applied evokes passive hyperpolarization of the membrane; this lowers P_K, which, in turn, increases resistance and,
consequently, increases the voltage drop on the membrane, and
so on. When P_K is lowered to a certain stationary value the HR
reaches a steady plateau, during which the membrane resistance
increases by 2.5-3.5 times by comparison with the initial resistance of a depolarized membrane. This resistance is usually
close in magnitude to the resistance of the resting membrane of
the same node in Ringer's solution. Only during passage of a very
strong hyperpolarization current did evidence of a "breakdown"
appear in the membrane, as shown by the appearance of irregular
waves of potential and, ultimately, by a fall in membrane resistance and steep decline of the HR plateau. This membrane breakdown was reversible in the sense that a membrane left at rest for
some time recovered and could retain about the same resistance
as before the first destructive action.

An essential feature of the membrane breakdown was that it
always appeared after a short latent period, during which the membrane remained capable of holding a high voltage. The duration of
this period, other conditions being equal, was inversely proportional to the degree of membrane hyperpolarization.

The presence of Ca^{++} in the solution of KCl is an essential
condition for the generation of normal HRs. In a "calcium-free"
medium or, more precisely, in a KCl solution containing traces
of Ca^{++},[*] the membrane became unstable and did not hold voltages
significantly higher than the initial transmembrane potential difference at rest.

Breakdown of the membrane and complete collapse of the HR
plateau developed under these conditions in response to the threshold strength of current for producing HR (Fig. 102a). If the node

---

[*] Frankenhaeuser (1957) showed that a so-called calcium-free solution contains not
less than 0.01 mM Ca^{++} because of the presence of these ions as an impurity in water
and salts.

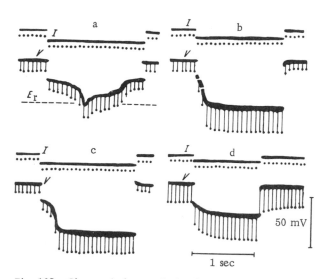

Fig. 102. Changes in hyperpolarization responses under the influence of Ca^{++} and La^{+++} ions: a) node kept in 114 mM KCl
"calcium-free" solution; b) addition of 10 mM $CaCl_2$ to the
medium; c) solution of 114 mM KCl + 10 mM $CaCl_2$ replaced
by one of 114 mM KCl + 0.1 mM $LaCl_3$; d) increase in $LaCl_3$
concentration to 10 mM. Top beam (I) shows applied hyperpolarizing current; bottom beam (V) shows changes in potential (Khodorov and Peganov, 1969).

was kept for a long time in a solution of 114 mM KCl and, in particular, if 2-4 mM EDTA (a compound binding Ca^{++}) was added to
it, after several repeated applications of hyperpolarizing current
the membrane often completely lost its ability to generate normal
HRs (Fig. 102a,b). Under these conditions addition of Ca^{++} to the
solution restored HRs (Fig. 102b).

The resistance of the Ranvier node membrane to breakdown
increases with an increase in the Ca^{++} concentration.

What is the nature of the breakdown phenomenon?

Breakdown can be obtained in nodes of Ranvier when kept not
only in KCl solution, but also in Ringer's solution of normal composition (Ooyama and Wright, 1961). Replacement of Cl^- by SO_4^{--}
in our experiments did not weaken the breakdown. Blocking the
potassium channels by means of TEA likewise had no effect on resistance of the membrane to breakdown.

It can be concluded from all these findings that the decrease in membrane resistance during strong hyperpolarization is due to a nonspecific increase in its ionic conductance, formally, to an increase in g_l.

Since breakdown phenomena become irreversible during excessively strong hyperpolarization (complete rupture of the membrane takes place) (Stämpfli and Willi, 1957), there is reason to suppose that they are based on some form of disturbance of the microstructure of the membrane, such as rupture of intermolecular bonds in lipoprotein complexes, leading to the appearance of new pores or ion-exchange areas and, as a consequence, to an increase in the ionic conductance.

The clear relationship between membrane resistance to breakdown and the concentration of Ca^{++} ions in the medium suggests that these ions stabilize the membrane structure linking phosphate and/or carboxyl groups of adjacent molecules of phospholipids and proteins together by means of bridges (Takata et al. 1966; Tasaki, 1967).

Considerable support for this suggestion can be found in the following facts.

a. Model experiments of Blaustein and Goldman (1966b) have shown that the affinity of divalent cations for one of the most important membrane phospholipids, phosphatidylserine, increases in the order: $Mg^{++} < Ca^{++} < Ba^{++} < Ni^{++}$. The stabilizing action of these ions on the membrane of the Ranvier node increased in exactly the same order in our experiments (Fig. 103).

b. The trivalent La^{+++} cation possess very high affinity for phosphatidylserine — hundreds of times higher than Ca^{++} (Rojas, Lettvin, and Pickard, 1966). The stabilizing action of La^{++} on the node membrane is correspondingly much stronger than the analogous action of Ca^{++} ions (Fig. 103).

Addition of La^{+++} to the KCl solution in a concentration of 0.1 mM had about the same stabilizing action on the node membrane as 10 mM Ca^{++} (Fig. 102c). An increase in La^{+++} concentration to 10 mM, however, made the membrane so stable that the voltage applied to it in order to produce breakdown had to be 3–5 times greater than in the case of action of 10 mM Ca^{++}.

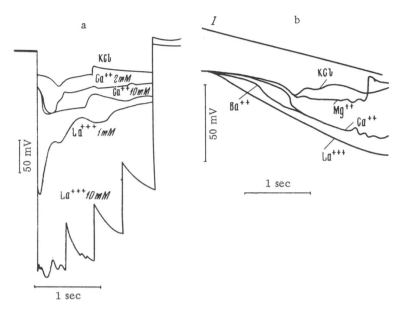

Fig. 103. Effect of polyvalent cations on the resistance of the membrane to breakdown. Experiments on the single Ranvier node. a) Effect of Ca^{++} and La^{+++} ions in different concentrations; breakdown produced by steady hyper-polarizing current. b) Effects of Mg^{++}, Ca^{++}, Ba^{++}, and La^{+++} ions taken in equimolar concentrations (5 mM); breakdown produced by linearly increasing hyperpolarizing current. In all cases KCl concentration in solution 114 mM (Khodorov and Peganov, 1969).

It is a very interesting fact that La^{+++} ions, despite the marked changes in the resistance of a depolarized (114 mM KCl) membrane, had virtually no effect on the magnitude of the membrane potential. In this respect their action differed from that of TEA, producing definite reduction of membrane depolarization.

c. Electron-microscopic investigations have shown that the presence of Ca^{++}, Ba^{++}, and, in particular, La^{+++} ions in solutions used for fixing tissues considerably stabilizes the structure of cell membranes. Lanthanum is definitely concentrated in the outer part of the membrane, giving it increased contrast in the electron microscopic image (Doggenweiler and Frenk, 1965).

On the basis of all these facts the temptation is very great to suggest that changes in ionic permeability of the excitable membrane on its depolarization and hyperpolarization are due to the effect of

the electrical field on binding of Ca^{++} ions or other polyvalent cations similar to Ca^{++} by the phospholipids of the membrane. Like
Karreman and Landahl (1952) and Koketsu (1965), one can assume
that the electrical field on the membrane changes the concentration
of free Ca^{++} ions in it, and in turn, this increases (during hyperpolarization) or decreases (during depolarization) the number of
phospholipid molecules at the gate of the ionic channels bound with
these ions.

Results showing the effect of La^{+++} ions on hyperpolarization
responses in the node of Ranvier (Fig. 102c,d), in the authors
opinion, are convincing evidence in support of this view.

In the joint investigation with E. M. Peganov cited above, the
author showed that La^{+++} ions not only promote an increase in resistance of the membrane during hyperpolarization (i.e., lower the
raised value of P_K), but they also cause a very considerable increase in the resting resistance (Fig. 101d). Hyperpolarization
causes a further increase in this effect (Fig. 101d), and after the
hyperpolarizing current is broken the membrane resistance returns only very slowly to its initial low level.* The time constant of this after-decrease in resistance lengthens with an increase in the La^{+++} concentration: in 0.5 mM La^{+++} it is about 50
msec, and in 2.5 and 10 mM La^{+++} it is 200 and 500 msec respectively. I cannot see how these facts can be explained without assuming that the increase in resistance produced by La^{+++} is due to
binding of these ions with the structural elements (most probably
with the phospholipids) of the membrane. Hyperpolarization increases this binding of La^{+++}, and since the lanthanum−phospho-
lipid complexes are very stable, they are split only very slowly
after the current is broken.

3. The parallel observed between the ability of the polyvalent
cations to stabilize the structure of the membrane (in the sense
indicated above) and the magnitude of the shifts in the g_{Na} versus
V curve along the voltage axis produced by these ions is of particular interest to the problem of excitability. According to Blau-

---

*It is clear from Fig. 101c that if Ca^{++} ions act after the end of hyperpolarization
the resistance will fall even below the initial level. This effect is due to the abolition of potassium inactivation, which is always present in a membrane depolarized
for a long time by K^+ ions (see Moore, 1967).

stein and Goldman (1968) Co^{++}, Cd^{++}, and Zn^{++} ions, in equimolar concentrations, produce an increase in φ_{Na} 3 or 4 times
greater than Ca^{++} ions. The effect of Ni^{++} and La^{+++}, on the other
hand, is 5 and 20–25 times respectively greater than the effect of
Ca^{++}. The action of Mg^{++} ions is weaker than that of Ca^{++}. In
other words, these cations are all arranged in the same order as
that obtained by Blaustein and Goldman (1966b) when comparing
their ability to bind with phospholipids. The special feature distinguishing the action of La^{+++} ions is that, besides an increase
in φ_{Na}, they also produce a marked decrease in the maximum
sodium and potassium conductances (Takata et al., 1966). The results cited above were obtained in experiments on crustacean axons.
They agree well with the results of the writer's comparative analysis of the effect of various polyvalent cations on the critical level
of depolarization of the node of Ranvier, and also with the results
obtained by Hille (1968a) on the same object, when comparing the
effects of Ca^{++} and Ni^{++} on the value of φ_{Na}. The difference between the sensitivity of the sodium and potassium channels to Ca^{++}
and H^{+}, discovered by Hille, is very important. The sodium channels were much more sensitive to changes in the concentration of
Ca^{++} in the medium, while potassium channels were more sensitive to changes in pH. Hille postulates the existence of charged
fixed groups, differing in their affinity for bivalent cations and H^{+},
close to the sodium and potassium channels (see also Hille, 1971).

Hypotheses on Conformational Changes

in the Membrane

The difficulties confronting the hypotheses that ionic carriers
or special ionic pores exist in the membrane led Tasaki and co-
workers (Tasaki and Singer, 1966; Tasaki, 1968) and other investigators to use a different approach to the analysis of excitation
phenomena based on the concept of phasic changes in the ion-exchange properties of the cell membrane.

According to Tasaki's hypothesis the excitable membrane is
a macromolecular complex of proteins and phospholipids. Each
of its subunits possesses ion-exchange properties and may exist
in one of two conformational states: a stable resting state and a
stable state of activity. In the former state the fixed negative
charges of the outer layer of the membrane, mainly carboxyl

groups, are occupied by bivalent cations forming stable complexes with lipoprotein "macromolecules." The latter state arises when the bivalent cations are displaced from these complexes by monovalent cations (Na^+, K^+). This displacement can take place either as the result of a very substantial increase in the K^+ ion concentration in the solution, or by the action of a depolarizing current on the membrane, leading to a marked increase in the K^+ ion flux from the axoplasm into the outer solution (Tobias, 1964). The action potential of the nerve or muscle fiber, in Tasaki's opinion, is nothing more than the expression of this phasic transition of the membrane macromolecules from the resting conformation into the conformation of activity. Tasaki links the end of excitation with the accumulation of cations, diffusing toward each other, taking place under these circumstances in and near the membrane.

An important argument in support of Tasaki's hypothesis is the following fact. Experiments with ion-exchange resins have shown that the substitution of divalent cations by monovalent ones is an exothermic process. Meanwhile, Hill and co-workers (Abbot, Hill, and Howarth, 1958) and Keynes, Ritchie, and Howarth (1965) showed that, contrary to the predictions of the theory of the concentration cell, during action potential generation heat is initially evolved, and only later absorbed.

It was further shown in Tasaki's laboratory that a sufficiently strong heat impulse, or a sudden increase in $[Ca]_0$ during the time of the plateau of the long action potential of the node of Ranvier, cuts short this plateau and rapidly restores the initial level of membrane polarization (the transition from the stable state of activity to the stable resting state) (Spyropoulos, 1961).

While disputing the existence of specific ionic channels in the excitable membrane, Tasaki, however, gives no convincing explanation of such important facts as the independent blocking of the early and late ionic currents by tetrodotoxin and tetraethylammonium, or the agreement between the equilibrium potentials for these currents and the calculated values of E_{Na} and E_K (see page 75), and so on.

Another variant of the "ion-exchange hypothesis" was suggested by Goldman (1964, 1965). Like Tasaki, Goldman considers that ion-exchange functions in the membrane are performed by anionic areas of the outer layer of the membrane which, in a resting state, are occupied by Ca^{++} ions. In Goldman's opinion, however, the leading role in this process is played, not by carboxyl groups, but by phosphate groups; he also postulates that the spatial orientation of the terminal groups of the phospholipids, which are flexible dipoles, can be modified by the electrical field in the membrane. A change in configuration of the dipole in turn affects the affinity of the phosphate groups for particular cations.

Goldman postulates the existence of three fundamental dipole configurations, the first (I) which binds (or absorbs) Ca^{++} ions preferentially, while the other two bind Na^+ (II) and K^+ (III). During action potential generation there is a successive change in the configuration of the dipoles in the direction from I to II and III.

In Goldman's idea, the phospholipid dipoles regulate only the admission of certain ions into the membrane phase. Their subsequent movement through the membrane is effected by concentration and electrical gradients. He does not rule out the possibility that such movement may also take place through special channels in the membrane (Goldman, 1965).

Blaustein and Goldman (1966a,b) used this hypothesis in an attempt to explain the differences between the action of certain local and general anesthetics on excitable structures and, in particular, the restorative action of Ca^{++} ions on electrical activity of nerve fibers treated with procaine (see page 117). Their basic idea was that the anesthetic competes with Ca^{++} ions for the bond with the phospholipid (Feinstein, 1964) in configuration I. The resulting phospholipid−anesthetic complex dissociates much more slowly than the phospholipid−Ca complex, as a result of which the Ca can take no part in ion-exchange reactions which usually develop on depolarization.

Unfortunately, in the concrete form in which it was proposed, Goldman's hypothesis has not been confirmed in experiments on artificial phospholipid membranes. Such membranes possess very low ionic permeability, which is virtually independent of the transmembrane potential difference. Electrical excitability arises in such membranes, as we shall see below, only when certain proteins are applied to them. In Goldman's hypothesis, however, the role of the protein components of the membrane in the mechanism of the changes in ionic permeability is totally disregarded.

Nevertheless, the underlying idea (which is common to both Tasaki's and Goldman's hypotheses) of the role of conformational changes in the mechanism of activation of the ionic permeabilities of the membrane during excitation is a very tempting one, and it has recently attracted an increasing number of supporters (Blumenthal, Changeux, and Lefevre 1970; Changeux et al., 1970, and others). Some very interesting new results, from this point of view, have recently been obtained: changes during the action potential generation in certain optical properties of the membrane, notably light scattering, birefringence (Cohen et al., 1968, 1970), and fluorescence (Tasaki et al., 1971). Unfortunately these results still remain very difficult to interpret (see Cohen et al., 1971).

Fishman, Khodorov, and Volkenstein (1970) attempted to make a quantitative analysis, from the point of view of these concepts, of the effects of the electrical field and of $[Ca]_0$ on activation of g_{Na}. The investigation showed that the initial slope of the exponential part of the g_{Na} versus E curve (see page 266), the gradient of this curve at the point of inflection, and its displacement with changes in $[Ca]_0$ (page 279) can be described quantitatively if it is assumed that activation of the sodium channels is associated with movement of the charge $n\delta \geq 4$ (where n is the valence and δ the fraction of E influencing movement of the charge) in the membrane, and that this movement leads to desorption of the Ca^{++} ions. The suggested model provides a qualitative explanation also for the effect of $[Ca]_0$ on the kinetics of changes in g_{Na}.

Investigations of Mechanisms of the Changes in Ionic Permeability on Artificial Membranes

Precipitation Membranes

The work of Kovalev, Liberman, and Chailakhyan (1966), and Chailakhyan (1967) has shown that the ionic permeability of precipitation membranes is definitely related to the transmembrane potential difference. This enabled them to use these membranes as models for reproducing certain properties of the excitable cell membrane.

   To obtain precipitation membranes, these workers used various microporous structures: filter paper, writing paper, cello-
phane, collodion and Formvar film, and so on. The membrane
was placed between two salt solutions (A and B), of which A (for
example $BaCl_2$) contained a cation and B (for example K_2SO_4) an
anion giving a sparingly soluble precipitate when added together.
Deposition of this precipitate in the pores of the membrane increased its resistance, and under the influence of a hyperpolarizing
current (from A to B) it increased still more. Under certain conditions "hyperpolarization responses" actually were produced and
were very similar outwardly to those observed on the nerve fiber
membrane.

   The mechanism of this increase in resistance is considered
by the workers cited to be as follows.

   An electric field, if applied to a membrane, changes the concentration of cations and anions in equilibrium with the precipitate,
and this change in equilibrium leads to further precipitation (during
hyperpolarization of the membrane) or to transfer of the precipitate
into solution (on depolarization). As a result, the resistance of the
artificial membrane appears smaller during depolarization than
during hyperpolarization.

   Attempts were made to use these results to explain the phenomena of selective permeability of cell membranes. The logic of
the arguments was as follows. The principal anion of the intercellular medium is Cl^-, which gives readily soluble compounds
with the cations present in the blood. Calcium ions, on the other
hand, exist in the cell in a virtually bound state. This suggests that
the cytoplasm contains many anions which give sparingly soluble
compounds with bivalent cations. The intracellular anions, meeting
bivalent cations in the pores of the membrane, are precipitated
and this is an important condition for maintenance of the high resistance of the membrane, its excitability, and so on (Chailakhyan,
1967).

   However, during internal perfusion of giant axons the membrane
retains its excitability and its high transmembrane resistance for
a long time, even if the principal anion is Cl^-. It is unlikely that
mobile anions, enabling continuous regeneration of the precipitate
to take place in the membrane pores, are preserved in the thin
(about 5 μ) juxtamural layer of cytoplasm remaining in these axons.

The suggestion that Ca^{++} ions interact in the membrane, not with anions in the cytoplasm, but with fixed negatively charged groups of the phospholipids and proteins of the membrane itself, and that they exert their influence on permeability to monovalent cations through changes in their conformation, is thus more likely to be nearer the truth.

Phospholipid Membranes

Results obtained on artificial phospholipid membranes are of the greatest interest to the current discussion.

Numerous investigations (see the survey by Goldup, Ohki, and Danielli, 1970) have shown that artificial phospholipid bimolecular films are similar in some of their physical properties to natural cell membranes. Their thickness is of the order of 50 Å, their capacitance about 0.5 $\mu F/cm^2$,* and their permeability to water molecules is similar to that characteristic of cell membranes. At the same time, the ionic permeability of phospholipid membranes is very low, so that their resistance to a direct current is five orders of magnitude higher than the resistance of membranes of nerve and muscle cells.

The usual explanation of these facts is that the phospholipid molecules in an artificial membrane are very tightly compressed by Van der Waals forces, with the result that only long, narrow "pores" accessible to water molecules, but impermeable to ions, remain between these molecules (Ermishkin, Liberman, and Smolyaninov, 1968).

However, there are certain compounds which confer on phospholipid membranes the property of selective ionic permeability. For instance, it has been shown (Müller and Rudin, 1967; Lev and Buzhinskii, 1967; Andreoli, Tieffenberg, and Tosteson, 1967) that the introduction of a very small quantity of the large-ring antibiotic valinomycin into a phospholipid bimolecular membrane gives this membrane selective permeability to potassium ions.† Under

---

* According to Babakov (1968) the capacitance of membranes made from brain phospholipids is 0.3 ± 0.2 $\mu F/cm^2$.

† A similar effect of valinomycin has been shown on the potassium permeability of mitochondria (Chappel and Croft, 1965) and erythrocytes (Tosteson et al., 1968).

these conditions the relative permeabilities for monovalent cations form the following series:

$$P_{Li} : P_{Na} : P_{Rb} : P_K = 1 : 1.4 : 210 : 395 : 920.$$

The mechanism of action of valinomycin has not been fully studied. However, it is known that the antibiotic can penetrate into the lipids of the membrane and form complexes with K^+ ions. The question whether valinomycin plays the role of a mobile K^+ carrier in phospholipid membranes or whether it forms ion-exchange channels in them (the "relay race" model) along which these ions move under the influence of the concentration gradient and electric field is at present being vigorously debated (Chizmadzhev and Markin, 1970; Stark and Benz, 1971).

The phospholipid membrane, treated with valinomycin, does not exhibit the properties of electrical excitability. However, this does take place under the influence of a small quantity (10^{-10} g/ml) of proteinaceous material, referred to as EIM (excitability-inducing material) (Müller and Rudin, 1967a, 1968). These workers obtained it from the bacterium Aerobacter cloacae, but it is also present in liver, brain, and brewers' yeast. The active component of EIM is evidently a ribonucleoprotein complex.

EIM gave the sphyngomyelin membrane the property of cationic permeability, as the result of which the resistance of the membrane fell by more than four orders of magnitude and a potential difference ("resting potential") was established between the two compartments of the chamber containing K_2SO_4 solution in different concentrations. Thereupon the membrane exhibited the property of delayed rectification. If, immediately after EIM, protamine sulfate was added to one compartment of the chamber, the artificial membrane became capable of producing responses, outwardly indistinguishable from the action potentials of the cell membrane, to application of a depolarizing current (Fig. 104). They had a clearly defined threshold and were accompanied by absolute refractoriness. This artificially created excitable system also gave anode-break excitation and spontaneous repetitive discharges, the frequency of which was controlled by temperature and electrical polarization. The responses were reversibly blocked by 2% cocaine and could be modified by acridine and phenothiazine derivatives, including chlorpromazine in very low concentrations.

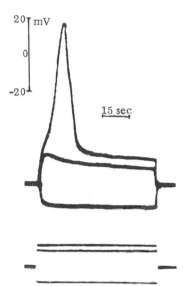

Fig. 104. Action potential obtained in artificial bimolecular phospholipid membrane (Müller and Rudin, 1967a).

Addition of Ca^{++} ions in millimolar concentrations blocked the repetitive discharges and increased the threshold.

Müller and Rudin emphasize the specificity of the components which they used. For example, histones, cholinesterase, and other basic proteins could not replace the protamine. In turn, protamine was inactive without EIM, binding with which (whether directly or through the phospholipids) was suppressed by an increase in the lecithin:sphyngomyelin ratio.

These workers interpret their results as follows. They consider that EIM forms channels in the phospholipid membrane with the following properties: they are selectively permeable to cations, capable of developing potentials in accordance with the Nernst equation, and they have a double "gate" mechanism corresponding to the two regions of negative resistance on the current–voltage curve (Müller and Rudin, 1963). Protamine evidently converts the cationic permeability of some of these channels into anionic without, however, changing the mechanism of operation of the "gates." This results in the formation of two populations of channels creating two opposite electromotive forces.

A successive change in the permeability of these channels under the influence of depolarization leads to the generation of action potentials. Müller and Rudin emphasize that this interpretation is compatible with the general demands of the Hodgkin—Huxley theory. However, in this particular case the anionic channels replace the sodium channels, and changes in ionic permeability are produced through a universal gate mechanism.

The similarity between the mechanism of action potential generation of the artificial membrane, postulated by Müller and Rudin, and that taking place in the algae Chara and Nitella, in which anionic permeability is also exchanged for cationic (Kishimoto, 1964).

Subsequent investigations by Bean et al. (1969) showed that after the addition of EIM the increase in conductance of the bimolecular membrane takes place in discrete steps, with an approximate magnitude of 4×10^{-10} mho, which supports the view that individual channels exist in the membrane. To study the properties of these channels, Ehrenstein, Lecar, and Nossal (1970) analyzed the spontaneous discrete steps of membrane resistance observed on the addition of very small quantities of EIM. As a result of their investigations these workers concluded that each of the ionic channels in a bimolecular membrane can be in two discrete states: open and closed. The proportion of the total number of channels open at any one time varies from zero to unity, as a function of potential. The negative resistance observed on membranes with a large number of channels is explained by a change in the ratio between the numbers of open and closed channels, and not by the conductance of each channel. The conductance of each channel in the open state was estimated to be $\sim 3 \times 10^{-10}$ mho.

It must be emphasized that EIM is not the only compound capable of conferring on phospholipid membranes the property of excitability. A similar effect, in principle, was obtained by Müller and Rudin (1968b) by treating such membranes with the cyclic polypeptide alamethicin. The chemical structure of this antibiotic has been much more fully studied than that of EIM, and this will probably facilitate the elucidation of the mechanism of its action.

The creation of artificial excitable membranes is unquestionably one of the most important achievements in the study of the molecular basis of cellular excitability.

Conclusion

The term electrical excitability defines the property of cell membranes to respond to changes of membrane potential by specific changes in ionic permeability. Most investigators consider that the electrical excitability of nerve fibers and of most muscle fibers is connected with the existence of separate and, evidently, spatially independent pathways or "channels" in the plasma membrane of these structures, through which Na^+ and K^+ ions can move down their respective electrochemical gradients.

Although the question of the physical nature of these channels remains unsolved ("carriers," "long pores," fixed-ion-exchange sites), their functional properties have now received intensive study.

Analysis of results obtained by the voltage-clamp technique shows that the sodium and potassium channels are qualitatively similar in their biophysical properties:

1. Their ionic selectivity is not absolute and, in many respects, it is similar to the relative ionic selectivity of glass electrodes;
2. Each channel can exist in three different states: at rest, when the channel is closed but can be activated by depolarization; active, when the channel is open; inactivated, when the channel is closed and cannot be activated by depolarization;

3. On depolarization of the membrane the channels are at first activated, but then change into a state of inactivation; hyperpolarization of the membrane, on the other hand, returns the channels to the state of rest either from the state of activity or from inactivation;

4. The time constants of activation and inactivation are bell-shaped functions of potential. However, despite the general qualitative similarity between these biophysical characteristics, their quantitative differences are very great.

For instance, the rate constants of sodium activation (α_m and β_m) are approximately one order of magnitude greater than the corresponding values of the rate constants of potassium activation (α_n and β_n) (Fig. 28). The time constants of inactivation are three orders of magnitude higher for the sodium channels than for the potassium channels. These differences in the rates of activation and inactivation of the ionic channels are exceptionally important from the functional point of view, for they determine the order of precedence whereby the inward sodium current (I_{Na}) increases first, and the outward potassium current (I_K) increases next, events lying at the basis of action potential generation (Fig. 43).

Besides the specific ionic channels, the electrically excitable membrane also contains nonspecific leak channels, the conductance of which (g_l) is virtually independent of the membrane potential. In nodes of Ranvier, with a high value of g_l and low P_K, the leak current plays an important role in spike generation.

An essential condition of action potential generation is an increase in the potential of the inside of the membrane E_r to the critical level E_c. During electrical stimulation of a nerve fiber this critical depolarization takes place as the result of summation of the passive (electrotonic) change in potential and the active subthreshold depolarization of the membrane (the local response), resulting from an increase in P_{Na} and I_{Na}.

The increase in P_{Na} during depolarization of the membrane is dependent on the potential and time. Its dependence on potential determines its regenerative character.

Critical depolarization of the membrane is necessary for the conversion of the local response into a spike because of the ex-

istence of three processes with a negative feedback function relative to regenerative depolarization: 1) an increase in inactivation
of the sodium channels inhibiting the growth of P_{Na} ; 2) activation
of the potassium channels with a resulting increase in I_K; and 3)
an increase in the outward leak current I_l .

Only at the critical potential (E_c) does the rate of increase of
the depolarizing sodium current (dI_{Na}/dt) begin to exceed the rate
of increase of the combined potassium current and leak current
[$d(I_K + I_l)/dt$], repolarizing the membrane, so that the local response is converted into a spike. During stimulation of a uniform
area of membrane by a short stimulus, ending before the action
potential begins, this condition is satisfied when I_{Na} becomes equal
in absolute magnitude to $I_K + I_l$.

During the action of a steady current (I_m), conversion of the
local response into a spike takes place before the equality $I_{Na} =
I_K + I_l$ is reached, since under these conditions regenerative de-
polarization takes place as the result of summation of I_{Na} and I_m .

It is customary to use the threshold potential (V_t) and the
threshold current (I_t) as indices of the excitability of a nerve fiber.
The value of V_t is determined by the difference between the critical potential and resting potential: $V_t = E_c - E_r$.

Different agents have different effects on E_r and E_c. Accordingly, depending on the direction and magnitude of the changes
in their values, V_t may rise, fall, or remain unchanged.

The resting potential (E_r) in nerve fibers with given concentration gradients of Na^+, K^+, and Cl^- ions is determined mainly by
the ratio between the coefficients of permeability (P_K/P_{Na}), each
of which, in turn, is a function of the number of open sodium and
potassium channels. The role of the electrogenic sodium pump in
the genesis of the resting potential is small under normal conditions.

Unlike E_r, the critical potential (E_c) depends principally on
the state of the sodium channel system, specifically: 1) on the total
number of sodium channels, which determines the maximum sodi-
um permeability ($\overline{P}_{Na}$); 2) the proportion of these channels in the
resting state (h_∞), i.e., free from inactivation; and 3) the reactivity of these channels (φ_{Na}).

Calculations using the Hodgkin–Huxley equations as modified by Frankenhaeuser for the node of Ranvier have shown that an isolated decrease in $\overline{P}_{Na}$ or h_∞, like an increase in φ_{Na}, leads to an increase in E_c. However, whereas a decrease in $\overline{P}_{Na}$ or h_∞, together with an increase in E_c, leads to a marked decrease in the amplitude (V_s) and, in particular, of the maximum gradient (V_{max}) of the spike, the main results of an increase in φ_{Na} are changes in E_c, and the changes in V_s and V_{max} are comparatively small (Figs. 31–42).

The reason for these differences is that when φ_{Na} is increased, the strength of I_{Na} is reduced only in the region of near-threshold changes in V, whereas when $\overline{P}_{Na}$ and h_∞ are reduced, I_{Na} falls for all values of the potential.

Tetrodotoxin, which blocks sodium channels, is an agent which specificially reduces $\overline{P}_{Na}$. Inactivation of the sodium channels can, however, be produced either by prolonged and/or strong depolarization of the membrane, or by treatment of the membrane with certain local (procaine, trimecaine, cocaine, etc.) and general anesthetics (barbiturates, ether, etc.).

The reactivity of the sodium channels, other conditions being the same, is a function of $[Ca]_0$: it rises with a decrease in $[Ca]_0$ and falls with an increase in $[Ca]_0$. The trivalent La^{+++} cations cause a simultaneous decrease in reactivity and in $\overline{P}_{Na}$.

The difference between sodium channels when inactivated and blocked by tetrodotoxin or La^{+++} is that the inactivated channels can be restored by strong hyperpolarization of the membrane or by an increase in $[Ca]_0$, whereas channels blocked by these agents cannot be restored under these conditions.

A change in the time constants of sodium activation (τ_m) or inactivation (τ_h) has little effect on the value of E_c, but it very considerably affects the rate of rise and fall of the action potential (Figs. 36 and 44).

The state of the potassium channel system likewise has little effect on the E_c level, which rises very slightly only in response to a very considerable increase in reactivity of the potassium system (to a decrease in φ_K). The potassium system has a substantial effect principally on the descending phase of the spike, prolonging it if φ_K is increased and shortening it if φ_K is reduced (Figs. 46 and 47).

Tetraethylammonium, which blocks potassium channels, causes a specific decrease in the maximum potassium conductance. The decrease in I_K resulting from its action leads to an increase in the duration of the descending phase of the action potential, which is particularly marked (with the appearance of a long plateau) in squid giant axons, in which I_K plays the leading role in the mechanism of membrane repolarization.

To understand the nature of the changes in excitability produced by the action of various factors on nerve and muscle fibers and cells, it is important to analyze the changes in critical potential associated with changes in the resting potential. During prolonged subthreshold depolarization of the membrane, E_c gradually rises through progressive sodium inactivation. In the case of prolonged hyperpolarization, E_c has a tendency to fall as a result of weakening of sodium inactivation. However, since the time constant of inactivation of the sodium channels is much greater than RC of the membrane, changes in E_c always lag behind the passive changes of membrane potential producing them. For this reason, during depolarization (i.e., an increase in E_r), the value of ($V_t = E_c - E_r$) at first falls, and then rises. During hyperpolarization, however, the value of V_t at first rises, and later falls; the lower the initial value of h_∞, the greater the fall.

Phenomena such as Pflüger's cateloctrotonus, enhancement of excitability in the phase of after-depolarization, and postsynaptic potentiation are based on a mechanism of an initial decrease in V_t during weak depolarization. The increase in E_c during depolarization, however, is the cause of the refractory state, or cathodic depression, and of Wedensky inhibition. The gradual increase in E_c during a slow rise of depolarization determines the phenomenon of accommodation to the stimulating current. The index of accommodation, namely, the critical slope of the linearly growing current, is directly dependent on the degree of initial inactivation of the sodium channels of the membrane $(1 - h_\infty)$ and inversely proportional to the time constant of inactivation (τ_h).

The ability of an excitable structure to give repetitive responses to a steady depolarizing current is also closely dependent on the value of h_∞, i.e., on the proportion of sodium channels which are free from inactivation at the resting potential.

Phenomena such as Pflüger's anelectrotonus and subnormality in the after-hyperpolarization phase are based on a mechanism of initial increase in V_t during hyperpolarization; postsynaptic inhibition is also partly due to hyperpolarization of the postsynaptic membrane.

Whereas in a uniform area of membrane E_c, like all the other parameters of the spike, is determined entirely by the state of the ionic channels, in the whole nerve fiber, with a cable-like structure, local currents generated between neighboring areas of the membrane likewise exert a considerable influence on these characteristics. For instance, in the continuous axon, because of the repolarizing effect of neighboring areas of the membrane, E_c is much higher, while the amplitude of the propagating action potential is lower, than in a uniform membrane.

The repolarizing action of local currents is particularly marked when conduction of the impulse into neighboring regions is interferred with by some means or other (for example, if the excitability of this region of the fiber is reduced or if the fiber is widened). Under these conditions the amplitude of the propagating spike falls as it draws nearer to this zone.

The decrease in amplitude of the spike before a zone of widening or branching of an axon increases the delay in conduction of the impulse in this region, which in turn leads to the generation of a retrograde depolarization wave. This prolongs the refractory state in the thin part of the fiber, just before it widens or branches, and it is the main cause of transformation of the rhythm of excitation (Wenckebach's period).

A retrograde wave of depolarization can arise in a geometrically homogeneous fiber if excitation is conducted from a segment with reduced spike amplitude (a region treated by anesthetics) to a zone in which the spike amplitude is normal.

Despite the detailed quantitative description of the kinetics of changes in membrane potential, ionic permeabilities, and membrane currents, the molecular mechanisms of these processes have not yet been explained.

None of the existing hypotheses regarding the nature of ionic permeability of excitable membranes can yet be regarded as having sufficiently solid experimental basis.

The main difficulties in the solution of this problem arise because we do not know enough about the structure of cell membranes or of the molecular transformations which take place during changes in membrane potential or the action of other stimuli.

The view which is most widely held at the present time is that electrically excitable membranes are pierced by channels which have two fundamental properties: selective permeability relative to ions of a certain type and the ability to alter this permeability in response to a change in membrane potential.

There is reason to suppose that these properties are due to the existence of special devices in each channel one of which determines the ionic selectivity of the channel while the other has the function of a "gating mechanism," opening or closing the channel in response to changes in the transmembrane potential difference.

An important role in this "gating mechanism" is played by Ca^{++} ions, which have affinity for the polar groups of the phospholipids of the membrane.

Addendum

While this book was in press some new important data appeared concerning the molecular mechanisms of the ionic permeability changes.

1. An agent was found which is able to increase specifically the resting sodium permeability of the nerve and muscle fibers. This agent is batrachotoxin(BTX)-steroidal alkaloid extracted from the skin secretion of the Colombian arrow poison frog. The increase of g_{Na} (to about 10% of $\overline{g}_{Na}$) leads to irreversible membrane depolarization. In squid giant axon the resting potential was eventually reversed by as much as 15 mV. The depolarization progressed more rapidly with internal application than with the external application of BTX to the membrane (550–1100 nM). External application of 1000 nM tetrodotoxin (TTX) completely restored the BTX depolarization. Despite the increase of resting P_{Na}, the BTX-poisoned membrane was still capable of undergoing a large permeability increase of normal amplitude upon depolarization provided the membrane potential was brought back to the original level. The action potential induced in such condition has a large and long-lasting negative after-potential (Narahashi, Albuquerque, and Deguchi, 1971). The presence of certain sulfhydryl and disulfide groups in the lobster axon membrane is essential for the action of BTX: previous treatment of the axon membrane with different sulfhydryl reagents resulted in a reduction of BTX depolarizing action (90%). These data strongly suggest the idea that BTX interacts with membrane protein.

The possible site of action of BTX is not the same as the site of action of TTX. This follows from the fact that pretreatment of the axon with sulfhydryl reagents does not change the effect of TTX.

2. The role of the protein components in the mechanism of membrane permeability changes was revealed quite distinctly in the experiments of Rojas and Armstrong (1971) on internally perfused squid axons (Dosidicus gigas). They found that the addition of 0.3–1.0 mg/ml pronase to the perfusion medium completely eliminates the sodium inactivation. Thus, after pronase treatment there is an abnormal sodium current that lasts as long as depolarization is maintained. TTX eliminates this current.

The results support the suggestion that inactivation depends on the intactness of a membrane protein.

3. Fox and Stämpfli (1971) have shown that ultraviolet radiation (wavelength 280 nm) induces an irreversible blockage of sodium channels. At 280 nm the sodium channels seem to be at least six times more sensitive than the potassium channels. The decrease of I_{Na} follows an exponential relation to the irradiation dose. In terms of target theory the single exponential relation of the dose−effect curve means that the irradiation effect is a one-hit event restricted to one target area. The volume of this area is estimated to be 200 $\overset{\circ}{A}{}^3$. The spectral sensitivity of the sodium channels to ultraviolet radiation is expressed by a curve which is very similar to the ultraviolet absorption curve of the protein (maximum at λ = 280 nm) (Fox, 1972).

The above-mentioned data of Rojas and Armstrong (1971) and of Fox (1972) favor the view of separate sodium and potassium pathways in excitable membranes.

4. The relative permeabilities of sodium channels to different metal and organic cations were studied in medullated nerve fibers (Hille, 1971, 1972). The measured reversal potential of the early ionic current and the Goldman equation were used to calculate permeability ratios. The main results are summarized in Table 4. Special experiments have shown that methyl and methylene groups render organic cations impermeant. This fact is explained on geometrical grounds assuming that the sodium channel is an oxygen-lined pore about 3 $\overset{\circ}{A}$ by 5 $\overset{\circ}{A}$ in cross section; one pair of oxygens is assumed to be an ionized carboxylic acid.

TABLE 4. Permeability Ratios
for All Measurably Permanent
Monovalent Cations in Sodium
Channel of Frog Node
(Hille, 1972a)

Ion	P_{ion}/P_{Na^+}
Sodium	1.0
Hydroxylamine	0.94
Lithium	0.93
Hydrazine	0.59
Thallium	0.33
Ammonium	0.16
Formamidine	0.14
Guanidine	0.13
Hydroxyguanidine	0.12
Potassium	0.086
Aminoguanidine	0.06

5. Protonation of an anionic group in the sodium channel at low pH (<6) induces a selective blockage of this channel. In the voltage-clamp experiments performed on Ranvier nodes Woodhull and Woodbury (1972) found that this blockage is voltage-dependent, being much stronger at E = 0 mV than at E = +160 mV. They assumed that H^+ ions can bind to sites within open sodium channels with voltage-dependent binding and unbinding rates obeying Eyring's rate theory. Positive internal potentials decrease H^+ binding by repelling H^+ ion binding sites. A new model has been proposed to account for this blockage of the sodium channel (Woodhull, 1972).

6. The ionic permeability of potassium channels of the Ranvier node membrane was investigated by Hille (1972). Relative permeabilities to test cations were calculated from reversal potentials using the Goldman equation. The main results are summarized in Table 5. The potassium channel, according to Hille, is much narrower than the sodium channel and therefore requires more fully dehydrated ions. The pore is physically too small for Cs^+ and too large for the close contact with Li^+ and Na^+ needed to offset the work of dehydration of these small ions.

TABLE 5. Permeability Ratio for
Monovalent Cations in Potassium
Channel (Hille, 1972b)

Ion	P_{ion}/P_{K^+}
Lithium	< 0.1
Sodium	< 0.05
Potassium	1.0
Thallium	2.80
Rubidium	2.96
Ammonium	0.13
Cesium	< 0.1

7. The specific late Ca^{++} channels in squid axon membrane
were discovered by Baker, Hodgkin, and Ridgway (1971). These
channels are sensitive to the blocking action of Mn^{++} ions and in-
sensitive to TTX, just like other calcium channels in the presynaptic ending, myocardial tissue, slow muscle, some molluscan
neurons, etc.

Bibliography

Abbot, B., Hill, A., and Howarth, J., 1958, Proc. Roy. Soc. B 148:149.

Aceves, J., and Machne, X., 1963, J. Pharmacol. Expt. Theor., 140:138.

Adelman, W. J., Dyro, F., and Senft, J., 1965a, J. Gen. Physiol., 48:521.

Adelman, W. J., Dyro, F., and Senft, J., 1965b, J. Cell. Comp. Physiol., Suppl. 2, 66:55.

Adelman, W. J., and Fok, J., 1964, J. Cell. Comp. Physiol., 64:429.

Adelman, W. J., and Palti, J., 1969, J. Gen. Physiol., 53:685.

Adelman, W. J., and Senft, J., 1966, J. Gen. Physiol., 50:279.

Adelman, W. J., and Taylor, R., 1961, Nature, 190:883.

Adrian, E., 1935, The Mechanism of Nervous Action [Russian translation], Moscow (Original Version, University of Pennsylvania Press, 1932).

Adrian, R., 1956, J. Physiol. (London), 133:631.

Adrian, R., 1960, J. Physiol. (London), 151:154.

Adrian, R., Chandler, W., and Hodgkin, A., 1970, J. Physiol. (London), 208:607.

Adrian, R., and Slayman, C., 1966, J. Physiol. (London), 184:970.

Agin, D., 1964, Nature, 201:625.

Albuquerque, E., 1972, Fed. Proc. 31:1133.

Andreoli, T., Tieffenberg, M., and Tosteson, D., 1967, J. Gen. Physiol., 50:2527.

Araki, T., 1960, Jap. J. Physiol., 10:518.

Araki, T., and Otani, T., 1955, J. Neurophysiol., 18:472.

Araki, T., and Otani, T., 1959, Jap. J. Physiol., 9:69.

Armstrong, C., 1966, J. Gen. Physiol., 50:491.

Armstrong, C., 1968, Nature, 219:1262.

Amstrong, G. M., and Binstock, L., 1965, J. Gen. Physiol., 48:859.

Arshavskii, Yu. I., Berkenblit, M. B., and Khodorov, B. I., 1969, Textbook of Physiology [in Russian], Vol. 1, Moscow.

Arshavskii, Yu. I., Berkenblit, M. B., Kovalev, S. A., Smolyaninov, V. V., and Chailakhyan, L. M., 1966. In Models of the Structural and Functional Organization of Some Biological Systems [in Russian], Moscow, p. 28.

Arshavskii, I. A., and Kurmaev, O. D., 1936, Fiziol. Zh. SSSR, 26:3.

Arvanitaki, A., and Chalosonitis, N., 1965, Compt. Rend. Soc. Biol., 159:1179.

Asano, T., and Hurlbut, W., 1958, J. Gen. Physiol., 41:1187.

Atwater, J., Bezanilla, F., and Rojas, E., 1969, J. Physiol. (London), 201:657.

Atwater, J., Bezanilla, F., and Rojas, E., 1970, J. Physiol. (London), 211:753.

Averbakh M. S., 1948, Vestn. Leningrad. Gos. Univ., 7:169.

Babakov, A. V., 1968, A Study of Bimolecular Phospholipid Models of Cell Membrane. Author's Abstract of Dissertation, Moscow.

Babskii, E. B., 1941, Transactions of the V. M. Bekhterev Institute of Brain Study [in Russian], No. 14, Leningrad.

Baker, P., 1965, J. Physiol. (London), 180:383.

Baker, P., 1966, Endeavour, 25:166.

Baker, P., 1971, XXV International Congress of Physiological Sciences, Vol. 8, p. 17.

Baker, P., Hodgkin, A., and Meves, H., 1964, J. Physiol. (London), 170:541.

Baker, P., Hodgkin, A., and Ridgway, E., 1971, J. Physiol. (London), 218:709.

Baker, P., Hodgkin, A., and Shaw, T., 1962, J. Physiol. (London), 164:355.

Baker, P., and Shaw, T., 1961, J. Marine Biol. Assoc. U.K., 41:855.

Baker, P., and Shaw, T., 1965, J. Physiol. (London), 180:424.

Bean, R., Shepherd, W., Chan, H., and Eichner, J., 1969, J. Gen. Physiol., 53:741.

Belyaev, V. I., 1963, Byull. Éksperim. Biol. i Med., No. 8, 24.

Belyaev, V. I., 1964a, Byull. Éksperim. Biol. i Med., No. 5, 3.

Belyaev, V. I., 1964b, Byull. Éksperim. Biol. i Med., No. 12, 13.

Bennett, M., 1967, Biophys. J., 7:151.

Bennett, M., Crain, S., and Grundfest, H., 1959, J. Gen. Physiol., 43:159.

Bergman, C., 1969, Seuil d'excitation et régimes d'activité du noeud de Ranvier. Thèses de la Faculté des Sciences d'Orsay, Université de Paris.

Bergman, C., and Stämpfli, R., 1966, Helv. Physiol. Pharmacol. Acta, 24:247.

Berkenblit, M. B., Kovalev, S. A., Smolyaninov, V. V., and Chailakhyan, L. M., 1966, in: Models of the Structural and Functional Organization of Some Biological Systems [in Russian], Moscow, p. 71.

Berkenblit, M. B., Vvedenskaya, N. D., Gnedenko, L. S., Kovalev, S. A. Kholopov, A.V., Fomin, S. V., and Chailakhyan, L. M., 1970, Biofizika, 15:81.

Berkenblit, M. B., Vvedenskaya, N. D., Gnedenko, L. S., Kovalev, S. A., Kholopov, A. V., Fomin, S. V., and Chailakhyan, L. M., 1971, Biofizika, 16:103.

Bernstein, J., 1902, Pflüg. Arch. Ges. Physiol., 92:521.

Bernstein, J., 1912, Elektrobiologie, Braunschweig.

Bezanilla, F., Rojas, E., and Taylor, A., 1970, J. Physiol. (London), 207:151.

Binstock, L., and Lecar, H., 1967, Abstracts of the Eleventh Meeting of the Biophysical Society, p. 19.

Binstock, L., and Lecar, H., 1969, J. Gen. Physiol., 53:342.

Bishop, G., 1928, Am. J. Physiol., 84:417.

Blair, E., 1932, J. Gen. Physiol., 15:709.

Blair, E., 1938, Am. J. Physiol., 123:455.

Blair, E., and Erlanger, J., 1932, Proc. Soc. Exp. Biol. (New York), 29:926.

Blair, E., and Erlanger, J., 1936, Am. J. Physiol., 114:2307.

Blaustein, M. P., 1968, J. Gen. Physiol., 51:293.

Blaustein, M. P., and Goldman, D., 1966a, J. Gen. Physiol., 49:1043.

Blaustein, M. P., and Goldman, D., 1966b, Science, 153:429.

Blaustein, M. P., and Goldman, D., 1968, J. Gen. Physiol., 51:279.

Blumenthal, R., Changeux, J., and Lefevre, R., 1970, Compt. Rend. Acad. Sci., 270D:389.

Bourguignon, G., 1923, La chronaxie chez l'homme, Paris.

Boyle, P. J., and Conway, E. J., 1941, J. Physiol. (London), 100:1.

Bradley, K., and Somjen, G., 1961, J. Physiol. (London), 156:75.

Brante, G., 1949, Acta Physiol. Scand., 18, Suppl. 63:1.

Braun, M., and Schmidt, R., 1966, Pflüg. Arch. Ges. Physiol., 287:56.

Brink, F., 1954, Pharmacol. Rev., 6:243.

Brinley, F., and Mullins, L., 1965, J. Neurophysiol., 28:526.

Burg, D., 1970, Pflüg, Arch. Ges. Physiol., 317:278.

Burke, W., Katz, B., and Machne, X., 1953, J. Physiol. (London), 122:588.

Burrows, T., Campbell, J., Howe, E., and Young, J., 1965, J. Physiol. (London), 179:39.

Bykhovskii, M. L., Korotkov, A. D., Rusanov, A. M., and Khodorov, B. I., 1966, Bio-
    fizika, 5:922.

Bykhovskii, M. L., Khodorov, B. I. Rusanov, A. M., and Korotkov, A. D., 1967, Bio-
    fizika, 13:529.

Caldwell, P., Hodgkin, A., Keynes, R., and Shaw, T., 1960, J. Physiol. (London),
    152:561.

Casteels, R., Drogman, G., and Hendrikx, H., 1071, J. Physiol. (London), 217:297.

Castillo, J., and Stark, I., 1952, J. Physiol. (London), 118:207.

Chagovets, V. Yu., 1896, Zh. Russk. Fiz.-Khim. Obshch., 28:431 and 657.

Chagovets, V. Yu., 1906, Outline of Electrical Events in Living Tissues in the Light
    of the Most Recent Physicochemical Theories [in Russian], Vol. 2, St. Petersburg.

Chagovets, V. Yu., 1937, Proceedings of the Sixth All-Union Congress of Physiologists
    [in Russian], Moscow, p. 355.

Chailakhyan, L. M., 1962, Biofizika, 5:639.

Chailakhyan, L. M., 1967, Elementary Properties of the Membrane and Electrical Be-
    havior of Excitable Tissues. Author's Abstract of Dissertation, Moscow.

Chandler, W., Hodgkin, A., and Meves, H., 1965, J. Physiol. (London), 180:821.

Chandler, W., and Meves, H., 1965, J. Physiol. (London), 180:788.

Chandler, W., and Meves, H., 1966, J. Physiol. (London), 186:121.

Chandler, W., and Meves, H., 1970, J. Physiol. (London), 211:623 and 707.

Changeux, J., Blumenthal, R., Kasai, M., and Podelski, T., 1970, Molecular Properties
    of Drug Receptors, New York, p. 197.

Chapman, R., 1966, J. Exp. Biol., 45:475.

Chappel, J., and Crofts, A., 1965, Biochem. J., 95:393.

Chizmadzhev, Yu. A., and Markin, V. S., 1970, Priroda, No. 6, 18.

Chweitzer, A., 1937, Ann. Physiol., 13:239 and 397.

Cohen, L., Keynes, R., and Hille, B., 1968, Nature, 218:438.

Cohen, L., Hille, B., Keynes, R., Landowne, D., and Rojas, E., 1971. J. Physiol.
    (London), 218:205.

Cohen, L., Keynes, R., Muralt, A., and Rojas, E., 1970, J. Phys. (Paris) 62:143.

Cole, K., 1968, Membranes, Ions, and Impulses, Berkeley, Calif.

Cole, K., 1928, J. Gen. Physiol., 12:37.

Cole, K., 1932, J. Gen. Physiol., 15:641.

Cole, K., 1941, J. Gen. Physiol., 25:29.

Cole, K., 1949, Arch. Sci. Physiol., 3:253.

Cole, K., 1958, J. Appl. Physiol., 12:129.

Cole, K., 1964, in: Problems in Biophysics. Proceedings of the First International
    Biophysical Congress [in Russian], Moscow.

Cole, K., and Curtis, H., 1936, Cold Spring Harbor Symp. Quant. Biol., 4:73.

Cole, K., and Curtis, H., 1938, J. Gen. Physiol., 22:37.

Cole, K., and Curtis, H., 1939, J. Gen. Physiol., 22:649.

Cole, K., and Moore, J., 1960, Biophys. J., 1:1.

Cole, K., Antoziewicz, H., and Rabinowitz, P., 1955, J. Soc. Indust. Appl. Math., 3:153.

Cole, K., Guttman, R., and Bezanilla, F., 1970, Proc. Nat. Acad. Sci. (Washington), 65:884.

Collwein, H., and Harreveld, A., 1966, J. Physiol. (London), 185:1.

Connelly, C., 1959, Rev. Mod. Phys., 31:475.

Connelly, C., 1962, Proceedings of IUPS, 22nd International Congress, Leiden, p. 600.

Connor, J., and Stevens, C., 1971, J. Physiol. (London), 213:21.

Conway, E., 1957, in: Metabolic Aspects of Transport Across Cell Membranes, Univ. Wisconsin Press, p. 73.

Conway, E., and Moore, J., 1945, Nature, 156:970.

Cooley, J., and Dodge, F., 1966, Biophys. J., 6:583.

Coombs, J., Curtis, D., and Eccles, J., 1957, J. Physiol. (London), 139:198.

Coombs, J., Curtis, D., and Eccles, J., 1959, J. Physiol. (London), 145:505.

Coombs, J., Eccles, J., and Fatt, P., 1955, J. Physiol. (London), 130:291.

Cremer, M., 1900, Z. Biol., 40:393.

Cremer, M., 1932, in: Handbuch der normalen und pathologischen Physiologie, Vol. 18, Berlin, p. 241.

Creutzfeldt, O., Lux, H., and Nacimiento, A., 1964, Pflüg, Arch. Ges. Physiol., 281:129.

Curtis, D., and Phillis, J., 1960, J. Physiol., 153:17.

Curtis, H., and Cole, K., 1938, J. Gen. Physiol., 21:757.

Curtis, H., and Cole, K., 1940, J. Cell. Comp. Physiol., 15:147.

Curtis, H., and Cole, K., 1942, J. Cell. Comp. Physiol., 19:135.

Dalton, J., 1958, J. Gen. Physiol., 41:529.

Davidov, V. A., 1942, Byull. Éksperim. Biol. i Med., 13:72.

Davson, H., and Danielli, J. F., 1943, Permeability of Natural Membranes, Cambridge Univ. Press.

Deck, K., 1958, Pflüg. Arch. Ges. Physiol., 266:249.

Deck, K., and Trautwein, W., 1964, Pflüg. Arch. Ges. Physiol., 280:63.

Derksen, H., 1965, Acta Physiol. Pharmacol. Neerl., 13:373.

Dettbarn, W., Higman, H., Rosenberg, P., and Nachmansohn, D., 1960, Science, 132:300.

Diecke, F., 1954, Z. Naturforsch., 96:713.

Dodge, F., 1961, in: A. M. Shanes (Editor), Biophysics of Physiological and Pharmacological Actions, Washington, p. 119.

Dodge, F., 1962, Proceedings of IUPS, 22nd International Congress, Leiden, p. 573.

Dodge, F., 1963, A Study of Ionic Permeability Changes Underlying Excitation in Myelinated Nerve Fibers of the Frog. Thesis, Rockefeller Institute, New York.

Dodge, F., and Frankenhaeuser, B., 1958, J. Physiol. (London), 143:76.

Dodge, F., and Frankenhaeuser, B., 1959, J. Physiol. (London), 148:188.

Doggenweiler, C., and Frenk, S., 1965, Proc. Natl. Acad. Sci. (Washington), 53:425.

Dray, S., and Sollner, K., 1956, Biochim. Biophys. Acta, 21:126.

DuBois Reymond, E., 1848, Untersuchungen über die thierische Electricität, Berlin.

Ebbecke, U., 1926, Pflüg. Arch. Ges. Physiol., 211:786.

Ebbecke, U., 1933, Ergebn. Physiol., 35:756.

Eccles, J. C., 1959, The Physiology of Nerve Cells [Russian translation], Moscow (Original Version, Oxford Univ. Press, London, 1957).

Eccles, J. C., 1966, The Physiology of Synapses [Russian translation], Moscow (Original Version, Springer, Berlin, 1964).

Eccles, J. C., Kostyuk, P. G., and Schmidt, R., 1962, J. Physiol. (London), 162:138.

Ehrenstein, G., and Gilbert, D., 1966, Biophys. J., 6:553.

Ehrenstein, G., Lecar, H., and Nossal, R., 1970, J. Gen. Physiol., 55:119.

Eichler, W., 1932, Z. Biol., 92:331.

Eichler, W., 1933, Z. Biol., 93:527.

Eisenman, G., 1964, in: Problems in Biophysics. Proceedings of the First International Biophysical Congress [in Russian], Moscow.

Eisenman, G., Sandlom, J., and Walker, J. L., Jr., 1967, Science, 155:965.

Engelman, T., 1870, Pflüg. Arch. Ges. Physiol., 3:247.

Erlanger, J., and Blair, E., 1931, Am. J. Physiol., 99:129.

Erlanger, J., and Blair, E., 1938, Am. J. Physiol., 121:431.

Ermishkin, L., Liberman, E., Smolyaninov, V., 1968, Biofizika, 13:205.

Fabre, P., 1936, Fiziol. Zh. SSSR, 21:208.

Fatt, P., and Ginsborg, B., 1958, J. Physiol. (London), 142:516.

Fatt, P., and Katz, B., 1951, J. Physiol. (London), 115:320.

Fatt, P., and Katz, B., 1953, J. Physiol. (London), 120:171.

Feinstein, M., 1964, J. Gen. Physiol., 48:357.

Fenn, W., 1936, Physiol. Rev., 16:450.

Fenn, W., and Haeg, L., 1942, J. Cell. Comp. Physiol., 19:37.

Fick, A., 1863, Beiträge zur vergleichenden Physiologie der irritablen Substanzen, Braunschweig.

Fishman, S., Khodorov, B., and Volkenstein, M., 1970, Biochim. Biophys. Acta, 225:1.

FitzHugh, R., 1960, J. Gen. Physiol., 43:867.

FitzHugh, R., 1961, Biophys. J., 1:446.

FitzHugh, R., 1962, Biophys. J., 2:11.

FitzHugh, R., 1966, J. Gen. Physiol., 49:989.

Fox, J., 1972, Veränderungen der spezifischen Ionenleitfähigkeit der Nervenmembran durch ultraviolette Strahlung, Dissertation.

Fox, J., and Stämpfli, R., 1971, Experientia, 27:1289.

Frank, K., and Fuortes, M. G. F., 1956, J. Physiol. (London), 131:424.

Frankenhaeuser, B., 1952, Cold Spring Harbor Symp. Quant. Biol., 17:27.

Frankenhaeuser, B., 1957a, J. Physiol. (London), 135:550.

Frankenhaeuser, B., 1957b, J. Physiol. (London), 148:671.

Frankenhaeuser, B., 1960, J. Physiol. (London), 151:491.

Frankenhaeuser, B., 1962a, J. Physiol. (London), 160:40.

Frankenhaeuser, B., 1962b, J. Physiol. (London), 160:46.

Frankenhaeuser, B., 1962c, J. Physiol. (London), 160:54.

Frankenhaeuser, B., 1963a, J. Physiol. (London), 169:424.

Frankenhaeuser, B., 1963b, J. Physiol. (London), 169:445.

Frankenhaeuser, B., 1965, J. Physiol. (London), 180:780.

Frankenhaeuser, B., and Hodgkin, A., 1957, J. Physiol. (London), 137:218.

Frankenhaeuser, B., and Huxley, A., 1964, J. Physiol. (London), 171:302.

Frankenhaeuser, B., and Moore, L. E., 1963, J. Physiol. (London), 169:438.

Frankenhaeuser, B., and Persson, A., 1957, Acta Physiol. Scand., 42, Suppl. 145, 45.

Frankenheauser, B., and Vallbo, A., 1965, Acta Physiol. Scand., 63:1.

Frankenhaeuser, B., and Widen, L., 1956, J. Physiol. (London), 131:243.

Fuortes, M., Frank, K., and Baker, M., 1957, J. Gen. Physiol., 40:735.

Gasser, H., 1935, Am. J. Physiol., 111:35.

George, E., and Johnson, E., 1961, Aust. J. Exp. Biol., 39:275.

Gerasimov, V.D., Kostyuk, P. G., and Maiskii, V. A., 1965a, Fiziol. Zh. SSSR, 60:703.

Gerasimov, V. D., Kostyuk, P. G., and Maiskii, V. A., 1965b, in: Protoplasmic Membranes and Their Functional Role [in Russian], Kiev, p. 24.

Gildemeister, Z., 1913, J. Biol., 62:358.

Goldman, D., 1943, J. Gen. Physiol., 27:37.

Goldman, D., 1964, Biophys. J., 4:167.

Goldman, D., 1965, J. Gen. Physiol., 48:75.

Goldman, L., and Binstock, L., 1969, J. Gen. Physiol., 54:735.

Goldstein, D., and Solomon, A., 1960, J. Gen. Physiol., 44:1.

Goldup, A., Ohki, S., Danielli, F., 1970 in Recent Progress in Surface Science, Vol. 3.

Golikov, N. V., 1950, Physiological Lability and Its Changes during Fundamental Nervous Processes [in Russian], Leningrad.

Gordon, H., and Welsh, J., 1948, J. Cell. Comp. Physiol., 31:395.

Gorman, A., and Marmor, M. J., 1970, J. Physiol. (London), 210:897, 919.

Graham, J., and Gerard, R., 1946, J. Cell. Comp. Physiol., 28:99.

Granit, R., and Scoglund, C., 1943, J. Neurophysiol., 6:337.

Granit, R., Kernell, D., and Shortness, G., 1963, J. Physiol. (London), 168:911.

Grilikhes, R. I., Rusanov, A. M., and Khodorov, B. I., 1967, Paper read at a Symposium on the Structure and Functions of Cell Membranes [in Russian], Pushchino.

Grundfest, H., 1932, J. Physiol. (London), 76:95.

Grundfest, H., 1964, in: Current Problems in Electrobiology [Russian translation], Moscow, p. 55.

Gul'ko, F., 1968, Biofiziki, 13:370.

Guttman, R., and Barnhill, R., 1966, J. Gen. Physiol., 49:1007.

Haapen, L., Kolmodin, G., and Scoglund, C., 1958, Acta Physiol. Scand., 43:315.

Hagiwara, S., Chichibu, S., and Naka, K., 1964, J. Gen. Physiol., 48:163.

Hagiwara, S., and Oomura, J., 1958, Jap. J. Physiol., 8:234.

Hagiwara, S., and Saito, N., 1959, J. Physiol. (London), 148:161.

Halidmann, C., 1963, Arch. Internat. Pharmacodyn., 146:1.

Hashimura, S., and Osa, T., 1963, Jap. J. Physiol., 13:3.

Hashimura, S., and Wright, E., 1958, J. Neurophysiol., 21:24.

Henneman, E., Somjen, G., and Carpenter, D., 1965a, J. Neurophysiol., 28:560.

Henneman, E., Somjen, G., and Carpenter, D., 1965b, J. Neurophysiol., 28:599.

Heppel, L. A., 1939, Am. J. Physiol., 127:385.

Hermann, L., 1879, Handbuch der Physiologie, Vol. 2, Leipzig.

Hermann, L., 1905, Pflüg. Arch. Ges. Physiol., 109:95.

Hill, A., 1936a, Proc. Roy. Soc. Ser. B., 119:305.

Hill, A., 1936b, Proc. Roy. Soc. Ser. B., 119:440.

Hille, B., 1966, Nature, 210:1220.

Hille, B., 1967a, A Pharmacological Analysis of the Ionic Channels of Nerve. Thesis, Rockefeller University, New York.

Hille, B., 1967b, J. Gen. Physiol., 50:1237.

Hille, B., 1968a, J. Gen. Physiol., 51:221.

Hille, B., 1968b, J. Gen. Physiol., 51:199.

Hille, B., 1970, in: Progress in Biophysics and Molecular Biology, 21:1.

Hille, B., 1971a, XXV International Congress of Physiological Sciences, Vol. 8, p. 41.

Hille, B., 1971b, J. Gen. Physiol., 58:599.

Hille, B., 1972a, J. Gen. Physiol., 59:637.

Hille, B., 1972b, Abstr. 16 Ann. Meeting.

Hodgkin, A. L., 1938, Proc. Roy. Soc. Ser. B., 126:87.

Hodgkin, A. L., 1948, J. Physiol. (London), 107:165.

Hodgkin, A. L., 1951, Biol. Rev., 26:339.

Hodgkin, A. L., 1954, J. Physiol, (London), 125:221.

Hodgkin, A. L., 1958, Proc. Roy. Soc. Ser. B., 148:1.

Hodgkin, A. L., 1964a, Science, 145:1148.

Hodgkin, A. L., 1964b, The Conduction of the Nervous Impulse, Liverpool Univ. Press.

Hodgkin, A. L., and Horowicz, P., 1959a, J. Physiol. (London), 145:405.

Hodgkin, A. L., and Horowicz, P., 1959b, J. Physiol. (London), 148·127.

Hodgkin, A. L., and Horowicz, P., 1960, J. Physiol. (London), 153:370.

Hodgkin, A. L., and Huxley, A. F., 1939, Nature, 144:710.

Hodgkin, A. L., Huxley, A. F., and Katz, B., 1949, Arch. Sci. Physiol., 3:129.

Hodgkin, A. L., Huxley, A. F., and Katz, B., 1952, J. Physiol. (London), 116:424.

Hodgkin, A. L., and Huxley, A. F., 1952a, J. Physiol. (London), 116:449.

Hodgkin, A. L., and Huxley, A. F., 1952b, J. Physiol. (London), 116:473.

Hodgkin, A. L., and Huxley, A. F., 1952c, J. Physiol. (London), 116:497.

Hodgkin, A. L., and Huxley, A. F., 1952d, J. Physiol. (London), 117:500.

Hodgkin, A. L., and Huxley, A. F., 1953, J. Physiol. (London), 121:403.

Hodgkin, A. L., and Katz, B., 1949, J. Physiol. (London), 108:37.

Hodgkin, A. L., and Keynes, R. D., 1954, Symp. Soc. Exp. Biol., 8:423.

Hodgkin, A. L., and Keynes, R. D., 1955a, J. Physiol. (London), 128:28.

Hodgkin, A. L., and Keynes, R. D., 1955b, J. Physiol. (London), 128:61.

Hodgkin, A. L., and Keynes, R. D., 1957, J. Physiol. (London), 138:253.

Hodgkin, A. L., and Rushton, W., 1946, Proc. Roy. Soc. B., 133:444.

Hoffman, B. F., and Cranefield, P. F., 1962, Electrophysiology of the Heart [Russian translation], Moscow. (Original version, New York, 1960).

Hoorweg, L., 1892, Pflüg. Arch. Ges. Physiol., 52:172.

Hoyt, R., 1963, Biophys. J., 3:399.

Hoyt, R., 1968, Biophys. J., 8:1074.

Hoyt, R., and Adelman, W., 1970, Biophys. J., 10:610.

Hutchinson, N., Cole, K., and Smith, R., 1970, J. Physiol. (London), 208:279.

Huxley, A. F., 1959, Ann. New York Acad. Sci., 81:221.

Huxley, A. F., 1964, Science, 145:1154.

Huxley, A., and Stämpfli, R., 1951a, J. Physiol. (London), 112:476.

Huxley, A., and Stämpfli, R., 1951b, J. Physiol. (London), 112:496.

Ichioka, N., 1957, Jap. J. Physiol., 7:20.

Ichioka, N., Uehara, J., and Kitamura, S., 1960, Jap. J. Physiol., 10:235.

Il'inskii, O. B., 1967, Problems in the Physiology of Sensory Systems, Mechanoreceptors [in Russian], Leningrad.

Ito, M., 1957, Jap. J. Physiol., 7:297.

Jenerick, H. P., 1956, J. Gen. Physiol., 39:773.

Jenerick, H. P., and Gerard, R., 1953, J. Cell. Comp. Physiol., 42:79.

Julian, F., Moore, J., and Goldman, D., 1962, J. Gen. Physiol., 45:1195.

Kandel, E., 1964, J. Gen. Physiol., 47:691.

Kandel, E., and Spencer, W., 1964, in: Current Problems in Electrobiology [Russian translation], Moscow, p. 241.

Kao, C., 1966, Pharmacol. Rev., 18:997.

Kao, C., and Nishiyama, A., 1965, J. Physiol. (London), 180:50.

Karreman, G., and Landahl, H., 1952, Cold Spring Harbor Symp. Quant. Biol., 17:293.

Kato, G., 1936, Cold Spring Harbor Symp. Quant. Biol., 4:202.

Katz, B., 1936, J. Physiol. (London), 88:239.

Katz, B., 1939, Electrical Excitation in Nerve, Oxford Univ. Press, London.

Katz, B., 1947, J. Physiol. (London), 106:66.

Katz, B., 1949, Arch. Sci. Physiol., 2:285.

Katz, B., 1950, J. Physiol. (London), 111:248.

Katz, B., and Thesleff, S., 1957, J. Physiol. (London), 137:267.

Kawai, J., and Sasaki, K., 1963, Jap. J. Physiol., 14:304.

Kerkut, C., and Thomas, R., 1965, Comp. Biochem. Physiol., 14:167.

Kernell, D., 1965, Acta Physiol. Scand., 65:65.

Kernan, R., 1962, Nature, 196:986.

Kernan, R., 1966, Proc. Roy. Irish Acad., Sect. B., 64:401.

Keynes, R. D., 1951, J. Physiol. (London), 114:119.

Keynes, R. D., and Lewis, P. R., 1951, J. Physiol. (London), 114:151.

Keynes, R. D., Ritchie, J. M., and Howarth, J., 1965, Proceedings of the 23rd International Congress of Physiologists, Tokyo, p. 85.

Khodorov, B. I., 1950, Uspekhi Sovr. Biol., 29:329.

Khodorov, B. I., 1962, Uspekhi Sovr. Biol., 54:333.

Khodorov, B. I., 1965a, in: Current Problems in the Physiology and Pathology of the Nervous System [in Russian], Moscow, p. 9.

Khodorov, B. I., 1965b, in: Protoplasmic Membranes and Their Functional Role [in Russian], Kiev, p. 46.

Khodorov, B. I., 1967, Biofizika, 6:1050.

Khodorov, B. I., and Belyaev, V. I., 1963a, Biofizika, 8:461.

Khodorov, B. I., and Belyaev, V. I., 1963b, Biofizika, 8:707.

Khodorov, B. I., and Belyaev, V. I., 1964a, Byull. Éksperim. Biol. i Med., No. 4, 3.

Khodorov, B. I., and Belyaev, V. I., 1964b, Tsitologiya, 6:680.

Khodorov, B. I., and Belyaev, V. I., 1964, Proceedings of the Tenth Congress of the I. P. Pavlov All-Union Physiological Society [in Russian], Vol. 3, pp. 11 and 311.

Khodorov, B. I., and Belyaev, V. I., 1965a, in: Biophysics of the Cell [in Russian], Moscow.

Khodorov, B. I., and Belyaev, V. I., 1965b, Biofizika, 10:625.

Khodorov, B. I., and Belyaev, V. I., 1966a, Biofizika, 11:108.

Khodorov, B. I., and Belyaev, V. I., 1966b, Biofizika, 11:288.

Khodorov, B. I., and Belyaev, V. I., 1967, Biofizika, 12:855.

Khodorov, B. I., Bykhovskii, M. L., Rusanov, A. M., and Korotkov, A. D., 1966, Biofizika, 11:1049.

Khodorov, B. I., Grilikhes, R. I., and Timin, E. N., 1970, Byull. Éksperim. Biol. i Med., No. 4, 24.

Khodorov, B. I., and Peganov, É. M., 1969, Biofizika, 14:3 [English translation: Neuroscience, Natl. Inst. Ment. Health, II, 30-40, 1969-1970].

Khodorov, B. I., and Timin, E. N., 1970, Biofizika, 15:503.

Khodorov, B. I., and Timin, E. N., 1971a, Biofizika, 16:495.

Khodorov, B. I., and Timin, E. N., 1971b, Neirofiziologiya, 3:316.

Khodorov, B. I., Timin, E. N., Pozin, N. V., and Shmelev, L. A., 1971, Biofizika, 16:95.

Khodorov, B. I., Timin, E. N., Vilenkin, S. Ya., and Gul'ko, F. B., 1969, Biofizika, 14:304.

Khodorov, B. I., Timin, E. N., Vilenkin, S. Ya., and Gul'ko, F. B., 1970, Biofizika, 15:140.

Khodorov, B. I., and Vornovitskii, E. G., 1967, Byull. Éksperim. Biol. i Med., No. 11, 34.

Kirzon, M. V., and Chepurnov, S. A., 1963, Trudy Moskovsk. Obshch. Estestvoisp. Prirody, Otdel. Biol., No. 9, p. 212.

Kishimoto, V., 1964, Jap. J. Physiol., 14:515.

Kita, H., 1965, Jap. J. Physiol., 16:70.

Kitamura, S., 1961, Jap. J. Physiol., 11:410.

Knutsson, E., 1964, Acta Physiol. Scand., 61: Suppl. 224:1.

Koketsu, K., 1965, Proceedings of IUPS, 23rd International Congress, Vol. 14, Tokyo, p. 521.

Koketsu, K., Cerf, A., and Nishi, S., 1959, J. Neurophysiol., 22:177.

Koketsu, K., Kitamura, R., 1964, Am. J. Physiol., 207:509.

Kolmodin, G., and Scoglund, C., 1958, Acta Physiol. Scand., 44:11.

Kompaneets, A. S., and Gurovich, V. T's., 1966, Biofizika, 11:913.

Koppenhöfer, E., 1964, Pflüg, Arch. Ges. Physiol., 279:36.

Koppenhöfer, E., 1966, Pflüg, Arch. Ges. Physiol., 289:9.

Koppenhöfer, E., 1967, Pflüg, Arch. Ges. Physiol., 293:34.

Koppenhöfer, E., and Schmidt, H., 1968, Pflüg. Arch. Ges. Physiol., 303:150.

Koppenhöfer, E., and Vogel, W., 1969, Pflüg. Arch. Ges. Physiol., 313:361.

Korn, D., 1966, Science, 153:1491.

Kostyuk, P. G., 1960, Microelectrode Techniques [in Russian], Kiev.

Kostyuk, P. G., and Semenyutin, I. P., 1961, Biofizika, 6:448.

Kovalev, S. A., Liberman, E. A., and Chailakhyan, L. M., 1966, Biofizika, 11:621.

Kries, J. von, 1884, Archives de Physiol., p. 337.

Kruyt, H., 1949, Colloid Science, Vol. 11, Reversible Systems, New York, pp. 227 and 299.

Kugelberg, E., 1944, Acta Physiol. Scand., 8, Suppl. 24.

Kurella, G. A., 1960, Biofizika, 5:260.

Kurella, G. A., and Liang Sing-tun, 1965, Biofizika, 10:72.

Kuryama, H., Osa, T., and Toida, N., 1967, J. Physiol. (London), 191:225.

Kvasov, D. G., 1937, Uspekhi Sovr. Biol., 7:1.

Kvasov, D. G., 1949, Vrach. Delo, 7:645.

Lapique, L., 1907, Compt. Rend. Soc. Biol., 62:925.

Lapique, L., 1908, Compt. Rend. Soc. Biol., 64:336.

Lapique, L., 1926, L'excitabilité en fonction du temps, Paris.

Lapique, L., 1936, Fiziol. Zh. SSSR, 21:1011.

Latmanizova, L. V., 1947, Vestn. Leningrad. Gos. Univ., 6.

Latmanizova, L. M., 1959, Techniques of Investigation of Accommodation of Excitable Tissues [in Russian], Moscow.

Lev, A. A., 1962, in: Fundamental Problems in the Evolutionary Physiology of the Central Nervous System [in Russian], Kiev, pp. 40-51.

Lev, A. A., 1964, in: Current Problems in Electrophysiological Research [in Russian], pp. 102-114.

Lev, A. A., and Buzhinskii, É. P., 1967, Tsitologiya, 9:102.

Liberman, E. A., 1963, A Study of the Mechanism of Biopotential Generation and Distribution of Ions between the Cell and Surrounding Medium. Author's Abstract of Dissertation, Moscow.

Liberman, E. A., and Chailakhyan, L. M., 1963, Tsitologiya, 5:311.

Liberman, E. A., Pronevich, L. M., Topaly, V. P., and Chailakhyan, L. M., 1967, Biofizika, 12:450.

Liberman, E. A., and Tsofina, L. M., 1960, Proceedings of the Third Conference on Electrophysiology of the Nervous System [in Russian], Kiev, p. 242.

Liesse, A., 1938, Compt. Rend. Soc. Biol., 128:1193.

Ling, G., 1960, J. Gen. Physiol., 43:149.

Ling, G., and Gerard, R., 1949, J. Cell. Comp. Physiol., 34:383.

Ling, G., and Gerard, R., 1950, Nature, 165:113.

Lowenstein, W. R., 1964, in: Current Problems in Electrobiology [Russian translation], Moscow.

Lorente de Nó, R., J. Cell. Comp. Physiol., 1944, 24:85.

Lorente de Nó, R., 1947, A Study of Nerve Physiology. Studies of the Rockefeller Institute for Medical Research, Parts 1-2, p. 132.

Lorente de Nó, R., Vidal, F., and Larremendi, L., 1957, Nature, 179:737.

Lucas, K., 1907, J. Physiol. (London), 35:310.

Lullies, H., 1930, Pflüg. Arch. Ges. Physiol., 225:69.

Lüttgau, H., 1953, Z. Naturforsch., 86:263.

Lüttgau, H., 1958, Pflüg. Arch. Ges. Physiol., 267:331.

Lüttgau, H., 1960, Pflüg. Arch. Ges. Physiol., 271:613.

Magnitskii, A. N., 1938, Byull. Éksperim. Biol. i Med., 5:466.

Magnitskii, A. N., and Muzheev, V. A., Trudy Fiziol. Otdel. Gos. Timiryazev. Inst. (Moscow), 77.

Maiskii, V. A., 1963a, Biofizika, 8:588.

Maiskii, V. A., 1963b, Electrical Characteristics of the Protoplasmic Membranes of Muscle Fibers and Nerve Cells in Different Functional States. Author's Abstract of Dissertation, Kiev.

Makarov, P. O., 1939, Trudy Leningrad. Obshch. Estestvoisp. (Leningrad), No. 1, 67.

Makarov, P. O., 1947, Problems in the Microphysiology of the Nervous System [in Russian], Moscow.

Makarov, P. O., 1949, Gagrskie Besedy (Tbilisi), Vol. 1.

Manery, J., and Ball, W., 1941, Am. J. Physiol., 132:215.

Markin, V. S., and Pastushenko, V. F., 1969, Biofizika, 14:316.

Marmont, G., 1949, J. Cell. Comp. Physiol., 34:351.
Medvedenskii, V. M., 1940, Byull. Éksperim. Biol. i Med., 10:47.
Meves, H., 1963, Pflüg. Arch. Ges. Physiol., 278:273.
Meves, H., and Weymann, D., 1963, Pflüg. Arch. Ges. Physiol., 275:357.
Michalov, V., Zachar, J., and Kostolansky, E., 1966, Physiol. Bohemoslov., 15:307.
Moore, J., 1965, J. Gen. Physiol., 48: Suppl. 11-7
Moore, J., 1967, J. Physiol. (London), 193:433.
Moore, J., Narahashi, T., and Shaw, T., 1967, J. Physiol. (London), 188:99.
Moore, J., Narahashi, T., and Ulbricht, W., 1964, J. Physiol. (London), 172:163.
Mosher, H., Fuhrman, F., Buchwald, H., and Fisher, H., 1964, Science, 144:1100.
Mozhaeva, G. N., Mozhaev, G. A., and Skopicheva, V. I., 1966, Biofizika, 11:462.
Mozhaeva, G. A., and Naumov, A. P., 1970, Nature, 228:164.
Müller, P., 1958, J. Gen. Physiol., 42:1013.
Müller, P., and Rudin, D., 1963, Theorct. Biol., 4:268.
Müller, P., and Rudin, D., 1967a, Nature, 213:603.
Müller, P., and Rudin, D., 1967b, Biochem. Biophys. Res. Commun., 26:298.
Müller, P., and Rudin, D., 1968a, J. Theoret. Biol., 18:222.
Müller, P., and Rudin, D., 1968b, Nature, 217:713.
Müller, P., Rudin, D., Tien, H., and Wescott, W., 1964, in: Recent Progress in Surface
    Science, Vol. 1, New York, p. 379.
Mullins, L., 1959, J. Gen. Physiol., 42:817.
Mullins, L., 1960, J. Gen. Physiol., 43:105.
Mullins, L., 1964, in: Current Problems in Electrobiology [in Russian], Moscow, p. 39.
Mullins, L., 1968, J. Gen. Physiol., 52:550.
Mullins, L., Adelman, W., and Sjodin, R., 1962, Biophys. J., 2:257.
Mullins, L., and Awad, M., 1965, J. Gen. Physiol., 48:761.
Muzheev, V. A., Sviderskaya, T. A., and Shitova, L., 1937, Arkh. Biol. Nauk, 45:93.
Nachmansohn, D., 1959, Chemical and Molecular Basis of Nerve Activity, New York.
Nakajima, S., 1966, J. Gen. Physiol., 49:629.
Nakamura, J., Nakajima, S., and Grundfest, H., 1965, J. Gen. Physiol., 48:985.
Narahashi, T., 1963, J. Physiol. (London), 169:91.
Narahashi, T., 1964, J. Cell. Comp. Physiol., 64:73.
Narahashi, T., Albuquerque, E., and Deguchi, T., 1971, J. Gen. Physiol., 58:54.
Narahashi, T., Anderson, N., and Moore, J., 1966, Science, 153:765.
Narahashi, T., Anderson, N., and Moore, J., 1967a, J. Gen. Physiol., 50:1413.
Narahashi, T., and Moore, J., 1968, J. Gen. Physiol., 52:550.
Narahashi, T., Moore, J., and Poston, R., 1967b, Science, 156:976.
Narahashi, T., Moore, J., and Scott, W., 1964, J. Gen. Physiol., 47:965.
Nasonov, D. I., 1949, "Bioelectrical Potentials," Gagrskie Besedy (Tbilisi), 1:1.
Nasonov, D. N., 1959, The Local Reaction of the Protoplasm and Spreading Excita-
    tion [in Russian], Leningrad.
Nasonov, D. N., and Alcksandrov, V. Ya., 1944, Uspekhi Sovr. Biol., 17:1.
Nastuk, L., and Hodgkin, A., 1950, J. Cell. Comp. Physiol., 35:39.
Nelson, P., 1958, J. Cell. Comp. Physiol., 52:127.
Nernst, W., 1908, Pflüg. Arch. Ges. Physiol., 122:275.
Nicholls, J., 1956, J. Physiol. (London), 131:1.

Niedergerke, R., 1953, Pflüg. Arch. Ges. Physiol., 258:108.

Nikol'skii, B. L., 1937, Zh. Fiz. Khimii SSSR, 7:587.

Noble, D., 1962, J. Physiol. (London), 160:317.

Noble, D., 1966, Physiol. Rev., 46:1.

Noble, D., and Stein, R., 1966, J. Physiol. (London), 187:129.

Ooyama, H., and Wright, E., 1961, Am. J. Physiol., 200:209.

Oikawa, T., Spyropoulos, C. S., Tasaki, I., and Teorell, T., 1961, Acta Physiol. Scand., 52:195.

Ostwald, W., 1890, Z. phys. Chem., 6:71.

Overton, E., 1902, Pflug. Arch. Ges. Physiol., 92:346.

Paoloni, L., 1963, in: Molecular Biology [Russian translation], Moscow (1963).

Parrack, H., 1940, Am. J. Physiol., 130:481.

Pastushenko, V. F., Markin, V. S., and Chizmadzhev, Yu. A., 1969, Biofizika, 14:883.

Perna, P. Ya., 1912, Transactions of the Physiological Laboratory, University of St. Petersburg [in Russian], Vol. 6, p. 3.

Pflüger, E., 1859, Untersuchungen über die Physiologie des Elektrotonus.

Pickard, W., Lettvin, J., Moore, J., Takata, M., Pooler, J., and Bernstein, T., 1964, Proc. Nat. Acad. Sci. (Washington), 52:1177.

Poussart, D., 1971, Biophys. J., 11:211.

Pryanishnikova, N. T., 1961, Dokl. Akad. Nauk SSSR, 141:1228.

Pryanishnikova, N. T., and Pchelin, V. A., 1959, Dokl. Akad. Nauk SSSR, 126:1358.

Pumphrey, R., Schmidt, O., and Young, J., 1940, J. Physiol. (London), 98:47.

Rall, W., 1962, Ann. New York Acad. Sci., 96:1071.

Rall, W., 1964, Proceedings of the First International Biophysical Congress, Moscow, 1961 [in Russian], p. 122.

Rashevsky, N., 1932, Protoplasma, 20:42.

Ritchie, J., 1961, in: Biophysics of Physiological and Pharmacological Actions, pp. 165-182.

Ritchie, J., and Straub, R., 1957, J. Physiol. (London), 136:80.

Ritter, cited by Dubois Reymond, E., 1848.

Robertson, R. N., 1964, in: Current Problems in Electrobiology [Russian translation], Moscow, p. 13.

Rojas, E., 1965, Proc. Nat. Acad. Sci. (Washington), 53:306.

Rojas, E., and Armstrong, C., 1971, Nature New Biol., 229:177.

Rojas, E., and Atwater, J., 1967a, Proc. Nat. Acad. Sci. (Washington), 57:1350.

Rojas, E., and Atwater, J., 1967b, Nature, 215:850.

Rojas, E., Bezanilla, F., and Taylor, R., 1970, Nature, 225:747.

Rojas, E., Lettvin, J., and Pickard, W., 1966, Nature, 209:886.

Rojas, E., and Luxoro, M., 1963, Nature, 199:78.

Rosenberg, H., 1935, J. Physiol. (London), 84:50.

Rosenblueth, A., 1952, Ergebn. Physiol., 24.

Rosenblueth, A., and del Pozo, E. C., 1942, Am. J. Physiol., 136:4.

Rosenblueth, A., and Luco, J. V., 1950, J. Cell. Comp. Physiol., 39:109.

Rosenblueth, A., and Ramos, G., 1951, J. Cell. Comp. Physiol., 39:109.

Rothenberg, M., 1950, Biochim. Biophys. Acta, 4:16.

Rushton, W. A. H., 1951, J. Physiol. (London), 115:101.

Rusinov, V. S., 1930, in: Collected Transactions of the Faculty of Physics and Mathematics, Leningrad University [in Russian], Leningrad.

Rusinov, V. S., 1934, Trudy Fiziol. Inst. Leningrad. Gos. Univ., 14:10.

Sasaki, K., and Oka, J., 1963, Jap. J. Physiol., 13:5.

Sasaki, K., and Otani, T., 1961, Jap. J. Physiol., 11:443.

Sato, M., 1951, Jap. J. Physiol., 2:255.

Schäfer, H., 1936, Pflüg. Arch. Ges. Physiol., 237:329.

Schmidt, H., 1963, Pflüg. Arch. Ges. Physiol., 278:4.

Schmidt, H., 1964, Ann. Univ. Saraviensis, 11:1.

Schmidt, H., and Stämpfli, R., 1966, Pflüg.Arch. Ges. Physiol., 287:311.

Schoepfle, G., and Erlanger, J., 1951, Am. J. Physiol., 167:134.

Schriever, E., 1932, Z. Biol., 93:123 and 249.

Scoglund, C., 1942, Acta Physiol. Scand., 4:Suppl. 12.

Segal, J., 1958, Nature, 182:1372.

Serkov, F. N., 1957, in: Current Problems in the Physiology of the Nervous and
    Muscular Systems [in Russian], Moscow, p. 455.

Shamarina, N. M., 1961, Fiziol. Zh. SSSR, 47:1046.

Shanes, A., 1954, Am. J. Physiol., 177:377.

Shanes, A., 1958, Pharmacol. Rev., 10:59.

Shanes, A., Freygang, W., Grundfest, H., and Amatiek, E., 1959, J. Gen. Physiol.,
    41:793.

Shapovalov, A. I., 1960, Biofizika, 3:270.

Shapovalov, A. I., 1964, Fiziol. Zh. SSSR, 50:444.

Shapovalov, A. I., 1965, Zh. Vyssh. Nervn. Deyat., 15:466.

Shapovalov, A. I., 1966, Cellular Mechanisms of Synaptic Transmission (from the
    Physiological and Pharmacological Aspects) [in Russian], Moscow.

Sjodin, R., 1959, J. Gen. Physiol., 42:983.

Sjodin, R., 1966, J. Gen. Physiol., 50:269.

Skou, J., 1957, Biochim. Biophys. Acta, 23:394.

Skou, J., 1965, Physiol. Rev., 45:596.

Solandt, D., 1936, Proc. Roy. Soc. Ser. B., 119:355.

Sollner, K., 1955, Arch. Biochim. Biophys., 54:129.

Sorokina, Z. A., 1964, Fiziol. Zh. SSSR, 50:040.

Spyropoulos, C., 1961, Am. J. Physiol., 200:203.

Spyropoulos, C., and Brady, R., 1959, Science, 129:1366.

Stallworthy, W., and Fensom, D., 1966, J. Physiol. Pharmacol., 44:866.

Stämpfli, R., 1952, Ergebn. Physiol., 47:70.

Stämpfli, R., 1954, Experientia, 10:508.

Stämpfli, R., 1958, Helv. Physiol. Pharmacol. Acta, 16:127.

Stämpfli, R., 1959, Ann. New York Acad. Sci., 81:127.

Stämpfli, R., and Nishie, K., 1956, Helv. Physiol. Pharmacol. Acta, 14:93.

Stämpfli, R., and Willi, M., 1957, Experientia (Basel), 13:297.

Stark, T., and Benz, R., 1971, J. Membrane Biol., 5:133.

Steinbach, H. B., and Spiegelman, S., 1943, J. Cell. Comp. Physiol., 22:187.

Straub, R., 1962, Proceedings of IUPS, 22nd International Congress, Leiden, pp. 598-
    599.

Takahashi, H., Murai, T., and Sasaki, T., 1960, Jap. J. Physiol., 10:280.

Takahashi, H., Usuda, S., and Ehara, S., 1962, Jap. J. Physiol., 12:545.

Takata, M., Moore, J., Kao, C., and Fuhrman, F., 1966, J. Gen. Physiol., 49:977.

Takata, M., Pickard, W., Lettvin, J., and Moore, J., 1966, J. Gen. Physiol., 50:461.

Takenaka, T., and Yamagishi, S., 1969, J. Gen. Physiol., 53:81.

Tarusov, B. N., 1960, Fundamentals of Biophysics and Biophysical Chemistry [in Russian], Moscow.

Tasaki, I., 1939, Am. J. Physiol., 125:367.

Tasaki, I., 1949, Biochim. Biophys. Acta, 3:498.

Tasaki, I., 1950, Jap. J. Physiol., 1:1.

Tasaki, I., 1953, Nervous Transmission, Thomas, Springfield.

Tasaki, I., 1956, J. Gen. Physiol., 39:377.

Tasaki, I., 1957, Conduction of the Nervous Impulse [Russian translation], Moscow.

Tasaki, I., 1959a, J. Physiol. (London), 148:306.

Tasaki, I., 1959b, Handbook of Physiology, Section 1, Neurophysiology, Washington, p. 75.

Tasaki, I., 1968, Nerve Excitation: A Macromolecular Approach, Charles C. Thomas, Springfield, Ill.

Tasaki, I., and Fujita, M., 1948, J. Neurophysiol., 11:311.

Tasaki, I., and Hagiwara, S., 1957, J. Gen. Physiol., 40:859.

Tasaki, I., and Mizuguchi, K., 1948, J. Neurophysiol., 11:295.

Tasaki, I., and Mizuguchi, K., 1949, Biochim. Biophys. Acta, 3:484.

Tasaki, I., Mizuguchi, K., and Tasaki, K., 1948, J. Neurophysiol., 11:305.

Tasaki, I., and Sakaguchi, M., 1950, Jap. J. Physiol., 1:7.

Tasaki, I., and Shimamura, M., 1962, Proc. Nat. Acad. Sci. (Washington), 48:1571.

Tasaki, I., and Singer, I., 1965, J. Cell. Comp. Physiol., Suppl. 2:66.

Tasaki, I., and Singer, I., 1960, Ann. New York Acad. Sci., 137:792.

Tasaki, I., Singer, I., and Takenaka, I., 1965, J. Gen. Physiol., 48:1095.

Tasaki, I., Singer, I., and Watanabe, A., 1965, Proc. Nat. Acad. Sci. (Washington), 54:763.

Tasaki, I., Singer, I., and Watanabe, A., 1966, Am. J. Physiol., 211:746.

Tasaki, I., Singer, I., and Watanabe, A., 1967, J. Gen. Physiol., 50:989.

Tasaki, I., and Takenaka, T., 1963, Proc. Nat. Acad. Sci. (Washington), 50:619.

Tasaki, I., and Takenaka, T., 1964, Proc. Nat. Acad. Sci. (Washington), 52:804.

Tasaki, I., and Takeuchi, T., 1942, Pflüg. Arch. Ges. Physiol., 245:764.

Tasaki, I. Mizuguchi, K., and Tasaki, K., 1948, J. Neurophysiol., 11:305.

Tasaki, I., Watanabe, A., and Hallett, M., 1971, Proc. Nat. Acad. Sci. (Washington), 68:938.

Tasaki, I., Watanabe, A., and Singer, I., 1966, Proc. Nat. Acad. Sci. (Washington), 56:1116.

Tasaki, I., Watanabe, A., and Takenaka, I., 1962, Proc. Nat. Acad. Sci. (Washington), 48:1177.

Tauc, L., 1955, J. Physiol. (Paris), 47:769.

Tauc, L., 1962a, J. Physiol. (Paris), 45:1077.

Tauc, L., 1962b, J. Physiol. (Paris), 45:1099.

Tauc, L., and Hughes, G., 1963, J. Gen. Physiol., 46:533.

Taylor, R., 1959, Am. J. Physiol., 196:1071.

Taylor, R., 1963, in: Physical Techniques in Biological Research, p. 219.

Teorell, T., 1949a, Arch. Sci. Physiol., 3:205.

Teorell, T., 1949b, Ann. Rev. Physiol., 11:545.

Teorell, T., 1964, in: Problems in Biophysics. Proceedings of the First International
    Biophysical Congress [in Russian], No. 22, Moscow, pp. 177-194.
Timin, E., 1971, The Spread of Stimuli in Nonuniform Excitable Structures, Author's
    Abstract, Moscow.
Timin, E. N., and Khodorov, B. I., 1971, Neirofiziologiya, 3:
Tobias, J., 1955, J. Cell. Comp. Physiol., 46:183.
Tobias, J., 1958, J. Cell. Comp. Physiol., 52:89.
Tobias, J., 1960, J. Gen. Physiol., 43:2 and 57.
Tobias, J., 1964, Nature, 203:13.
Tomita, T., 1966, J. Theoret. Biol., 12:216.
Tomita, T., and Wright, E., 1965, J. Cell. Comp. Physiol., 65:195.
Tosteson, D., Cook, P., Andreoli, T., and Tieffenberg, M., 1967, J. Gen. Physiol.,
    50:2513.
Tosteson, D., Cook, P., Andreoli, T., and Tieffenberg M., 1968, J. Gen. Physiol.,
    51:373.
Troshin, A. S., 1956, The Problem of Cell Permeability [in Russian], Moscow-Leningrad.
Troshin, A. S., 1964, Trudy Moskovsk. Obshch. Ispyt. Prirody, 9:7.
Truant, A., 1950, Fed. Proc., 9:321.
Tsofina, L. M., and Liberman, E. A., 1964, Biofizika, 9:242.
Uchizano, K., 1960, in: Electrical Activity of Single Cells, p. 97.
Uehara, J., 1960, Jap. J. Physiol., 10:3.
Uflyand, Yu. M., 1941, The Theory and Practice of Chronaximetry [in Russian],
    Leningrad.
Ukhtomskii, A. A., 1950, Collected Works [in Russian], Vol. 1, Moscow.
Ulbricht, W., 1963, Pflüg. Arch. Ges. Physiol., 277:270.
Ulbricht, W., 1964, J. Gen. Physiol., 40:110.
Ushakov, B. P., et al., 1959, Fiziol. Zh. SSSR, 39:218.
Ushiama, J., Koizumi, K., and Brooks, C., 1966, J. Neurophysiol., 29:1028.
Ussing, H., 1947, Nature, 160:262.
Ussing, H., 1949, Acta Physiol. Scand., 19:43.
Ussing, H., 1949, Physiol. Rev., 29:127.
Vallbo, A., 1964a, Acta Physiol. Scand., 61:413.
Vallbo, A., 1964b, Acta Physiol. Scand., 61:429.
Vasil'ev, L. L., 1925, Advances in Reflexology and the Physiology of the Nervous
    System [in Russian], p. 1.
Vasil'ev, L. L., Lapitskii, D. A., and Petrov, F. I., 1936, Fiziol. Zh. SSSR, 21:988.
Vasilevskii, V. M., 1946, Data on the Nervous Regulation of Activity of Warm-Blooded
    Animals and Man. Author's Abstract of Dissertation, Khar'kov.
Verveen, A., 1961, Fluctuation in Excitability, Research Report on Signal Transmis-
    sion in Nerve Fibers, Amsterdam.
Verveen, A., 1962a, Acta Morphol. Neerl. Scand., 5:79.
Verveen, A., 1962b, Acta Physiol. Pharmacol. Neerl., 10:294.
Verveen, A., 1962c, Acta Physiol. Pharmacol. Neerl., 11:
Verveen, A., 1962d, Proceedings of IUPS, 22nd International Congress, Vol. 2, Leiden,
    p. 788.
Verveen, A., and Derksen, H., 1968, Proc. IEEE, 56:906.

Villegas, R., Villegas, G., Blei, M., Herrera, F., and Villegas, J., 1966, J. Gen. Physiol.,
    50:43.
Vornovitskii, E. G., 1966, Byull. Éksperim. Biol. i Med., No. 10, p. 11.
Vornovitskii, E. G., and Khodorov, B. I., 1966, Byull. Éksperim. Biol. i Med., No. 6, p. 8.
Vornovitskii, E. G., and Khodorov, B. I., 1967, Byull. Éksperim. Biol. i Med., No. 9,
    p. 3.
Vorontsov, D. S., see Woronzow, D.
Vorontsov, D. S., and Shuba, M. F., 1966, The Physical Electrotonus of Nerves and
    Muscles [in Russian], Kiev.
Watanabe, A., Tasaki, I., Singer, I., and Lerman, L., 1967, Science, 155:95.
Weakly, L., 1969, J. Physiol. (London), 204:63.
Wedensky, N. E., 1901, Excitation, Inhibition, and Narcosis [in Russian], St. Petersburg.
Wedensky, N. E., 1922, Cited by Ukhtomskii, A. A., 1950.
Weidmann, S., 1955, J. Physiol. (London), 127:213.
Weiss, G., 1901, Compt. Rend. Soc. Biol., 53:253.
Werigo, B. F., 1883, Trudy SPb. Obshch. Estestvoispyt., 14:1.
Werigo, B. F., 1888, The Action of an Interrupted and Continuous Galvanic Current
    on Nerve (an attempt to explain the physiological events of electrotonus) [in
    Russian], St. Petersburg.
Werigo, B. F., 1901a, in: K. M. Bykov (Editor), The Physiology of the Nervous Sys-
    tem [in Russian], Vol. II, Moscow, 1952, p. 543.
Werigo, B., 1901b, Pflüg, Arch. Ges. Physiol., 84:547.
Whittembury, G., Sugino, N., and Solomon, A., 1960, Nature, 187:699.
Wolman, M., 1970, in: Recent Progress in Surface Science, Vol. 3, p. 261.
Woodhull, A., 1972, Ionic Blockage of Sodium Permeability in Voltage-Clamped Frog
    Nerve, Dissertation.
Woodhull, A., and Woodbury, J., 1972, Federation Meeting.
Woronzow, D. (Vorontsov, D. S.), 1924, Pflüg. Arch. Ges. Physiol., 203:300.
Woronzow, D. (Vorontsov, D. S.), 1925, Pflüg, Arch. Ges. Physiol., 207:279.
Wright, E., 1956, Proc. Soc. Exp. Biol. (New York), 93:318.
Wright, E., and Tomita, T., 1965, J. Cell. Comp. Physiol., 65:211.
Yamagiwa, K., 1956, Jap. J. Physiol., 6:167.
Young, J. Z., 1936, Quart. J. Microsc. Sci., 78:367.
Zachar, J., Zachareva, and Heneek, M., 1964, Physiol. Bohemoslov., 13:129.
Zencker, W., 1964, Z. Zellforsch., 62:531.
Zhukov, E. K., 1940, Trudy Leningrad. Obshch. Estestvoispyt., 68:1.
Zhukov, E. V., 1940a, Byull. Éksperim. Biol. i Med., No. 9, p. 49.
Zhukov, E. V., 1940b, Byull. Éksperim. Biol. i Med., No. 9, p. 50.

Index